Build a Jungle Zoo

Also by the author:

Keeping the Dwarf Sea Horse
Keeping Sea Horses
Sharks Morays and Treasure
The Salt Water Aquarium in the Home
Exploring the Reef
Jet Safari to Africa
The Marine Collector's Guide

Build a Jungle Zoo

Robert P. L. Straughan

South Brunswick and New York: A. S. Barnes and Company
London: Thomas Yoseloff Ltd

A. S. Barnes and Co., Inc.
Cranbury, New Jersey 08512

Thomas Yoseloff Ltd
108 New Bond Street
London W1Y OQX, England

Library of Congress Cataloging in Publication Data

Straughan, Robert P L
Build a jungle zoo.

Bibliography: p.
1. Open-air zoos—United States. 2. Animal introduction—United States. 3. Wildlife conservation—United States. I. Title.
QL76.5.U6S77 590'.744 79-37817
ISBN 0-498-07999-6

Unless otherwise credited, all photographs are by Robert P. L. Straughan.

Printed in the United States of America

To the wildlife conservationists, past and present,
whose vigilance and forethought have
preserved the wilderness areas and national parks
in our country and abroad. Long may they live!

Contents

Foreword

We are all guilty of the devastation of our wildlife. We cannot blame the zoo, the pet shop, the animal importer, or the individual who keeps an exotic pet. Nor can we blame the hunter who kills for meat or trophy. We are all to blame for the plight of our diminishing wildlife, whether we have been active in collecting or keeping animals or passive in simply not caring what has happened to the animals that share our planet with us.

We should hardly point a finger at another without first looking in the mirror, for here will be found an enemy of wildlife. Fur coats can only be made by the deaths of millions of animals. The collections in museums of bird and animal skins, which may number in the thousands or millions, have not increased our wildlife. In fact, the rarer the bird or animal, the more urgent the desire to collect it, which could in some instances result in the extinction of an entire species. The land developer who drains swamps and fells vast woodlands to build houses has destroyed wildlife habitat forever. Even the nature and conservation magazines cannot get away unscathed, for we must ask, how many million trees must be cut down just to print their thick magazines each year? Or perhaps we might ask, could not the money taken in from subscriptions and donations be better spent by putting out a slightly less pretentious magazine and using the extra funds to purchase new wilderness areas each year, before such lands are turned into housing lots? Nor can our own government escape a scolding finger, for it is a fact that government trappers still put out deadly poisonous baits to kill wolves and coyotes and in fact kill many other animals, even some that are on the verge of extinction. Yes, we are all guilty of wanton destruction of wildlife. Did you know that nearly half a *billion* animals are killed each year by cars alone? We need to save our wildlife. This book tells how.

Acknowledgments

The quest for information is constant, and the best source is from those with experience. In accumulating the mass of data for this book and for my proposed big-game preserve, I deluged many of the nation's most experienced animal handlers and zoo directors with questions that never seemed to end. In particular, I wish to thank Alton Freeman, who gave freely of his time on numerous occasions, never asking anything in return. I would also like to thank Bill Chase, Antonio Gomez, Warren Prince, Roland Lindemann, Fairfield Osborn, Theodore Reed, Henry Deens, Russ Smiley, William Conway, Robert Taliaferro, Ross Allen, Ivan T. Sanderson, Bill Haast, Carl Selph, Paul Zahl, V. Rider, Marlin Perkins, Richard Naegeli, Bruce Grant, Charles Schroeder, W. McMichael, Robert Austin, George Pournelle, the United States Department of Agriculture, the United States Plant Introduction Service, Dena Goluboff, editor, and others who helped with suggestions or criticism of my often grandiose ideas.

I also greatly acknowledge the following publishers, organizations, and individuals for their kind permission to reprint from published works and for the outstanding photographs that they supplied:

Busch Gardens, Tampa, Florida, Robert Taliaferro.

Field and Stream Magazine, Michael Frome, Conservation Editor.

Jungle Larry's African Safari, Naples, Florida, Lawrence E. Tetzlaff.

New Mexico Department of Game and Fish, Ladd S. Gordon, Robert H. Stewart, William S. Huey.

Pets, Supplies, Marketing Magazine, Duluth, Minnesota.

Rainbow Springs, Dunnellon, Florida, Richard L. Needleman.

Simian Society of America, Inc., Jan Tompkins.

World of Animals, Mesquite, Texas.

Introduction

There is little doubt that the wildlife of the world is in imminent danger. Many species border on the verge of extinction, which could occur within five years for some of them. Animal preserves in many of the countries of the world offer protection for wild creatures, but, unfortunately, some of the rarest animals are being killed within the preserves. It thus becomes apparent that we must do *more* than merely protect the animals in their native land, for in some instances this is insufficient to guarantee the positive survival of the species. Prime examples of this are the Asiatic lion, whose number has dwindled down to a few hundred in recent years, despite protection in the Gir forests in India. According to a recent on-the-spot estimate, there are now only 150 lions left, whereas just a few years ago there were five or six hundred. Religious beliefs that allow millions of disease-ridden cattle to wander unrestricted throughout the land have brought severe problems to India's once fabulous wildlife. Natives shoot the lions, which naturally prey on cattle that stray within their preserve. The huge Indian rhino, probably the most spectacular of all living animals today, is on the extreme danger list. Its number is down to five or six hundred animals, and although these live in protected areas, giant grasses and swamps make the area difficult to patrol. Animals are slaughtered and their bodies are never discovered. The Java rhino is in even worse shape. Only about thirty-five are said to survive. The list goes on and on.

In this book, the author presents many suggestions for increasing the wildlife throughout the world. He proposes giant outdoor jungle preserves in the southern areas of our country where the animals would roam free in huge wild areas. Herds of elephants, giraffes, zebras, and all the magnificent plains game including the cheetah would roam free. Only the very outer perimeter would be fenced in some areas. In others, there would be no fence at all; the animals would be free to roam the wilds and share the environment with our own native species. After all, the white man is not native and he is sharing the land with the Indians. Why not allow selective foreign animals to share our land?

It is a fact that in Texas there are now blackbuck and nilgai antelope living in the wilds, and some animal experts feel that there may well be more of these India natives in our own country than there are in all of India. In fact, a short while ago, a small herd of black buck were shipped from Texas *back* to West Pakistan to restock the animal in its native land. A few decades ago, the black buck lived by the tens of thousands in Pakistan but now are practically extinct there. Excessive hunting and the development of wild areas for human habitation have reduced the world's wildlife to its lowest level since man took over the earth. Polar bears and Siberian tigers are hunted by plane. Both are threatened species, but the giant Siberian tiger is now down to a few hundred animals in all the world. Even in the vast Amazon, the Indian now hunts with a high-powered rifle or shotgun, and many spotted cats that would have survived the bow are now falling victim to man's greed.

The author proposes giant jungles in America. He proposes new, outdoor-type zoos where the animals are no longer caged. He suggests the minizoo for cities and areas that cannot afford the larger compounds. He recommends that exotic animals purchased as pets be bought only as pairs so that an attempt can be made to raise them and propagate the species. But he does not merely suggest all of these things, he tells how they can actually be done based upon his own experiences in keeping animals and visiting most of the nation's zoos, national parks, and animal farms, as well as visiting the huge game preserves of Africa.

By creating vast wilderness areas in our own country and stocking them with exotic animals from Africa and Asia, we will not only be preserving the wildlife of the world but we will be saving our wilderness areas from becoming housing lots. In addition, a herd of elephants, the stately giraffe, or the thunderous rhino will add tremendous interest to our new national animal parks. Our present parks are practically devoid of animal life. You can see more animals in *one glance,* in Africa's big-game preserves than you can see in all of our own parks put together. In addition, only the wealthy can afford to go to Africa to see the spectacular big game in its natural habitat. Why not bring a little of Africa to the United States? It can be done.

Build a Jungle Zoo

1
The Case for the Exotic Pet

There is a modern-day cry from many conservationists to totally ban the importation of all exotic animals. This is to include monkeys, ocelots, jaguars, and practically every other kind of animal. Even birds are included. The reasons for the total ban are that the animals are mistreated, that they soon die in captivity, or that they die en route. Unfortunately, much of this is true. Not only do monkeys and birds die in wholesale lots at the airports or while in transit either from their country of origin or to distant points within the United States, but those that do arrive at their final destination are often so weakened that they have no chance of survival. Furthermore, in the case of gorillas, gibbons, chimps, and other members of this family, the young animals are collected usually by the horrible method of shooting the mother in order to wrest the baby from her. If those who keep monkeys as pets were aware of this, perhaps many of them would refuse to purchase another animal until this cruel and shameful practice is stopped.

The following article, reprinted from the *Simian,* the monthly publication of the Simian Society of America, an educational and humane organization dedicated to improving the welfare of monkeys and apes in captivity, sums up succinctly, the trials and tribulations which monkeys and apes endure before they end up as a pet or in a zoo.

Primate Perils in Trapping and Transportation
by Jan Tompkins

Monkeys are faced with many serious hazards from the moment they are trapped in the wild until they reach their final destination . . . be it zoo, house, or laboratory. Many real hurdles face the person who tries to transport a live, healthy simian from one part of the world to another.

Trapping

Trapping monkeys is easy. Humans have had thousands of years of practice. Methods vary a little but methods have changed only slightly over the centuries.

The usual African trap for Grivets or for Patas Monkeys resembles an inverted breadbasket weighted with a log or a stone and has a simple trigger device. Dry corn is commonly used for bait. For heavier, stronger animals like adult baboons, the hunters use self-triggering traps made from tough thornbush branches that have been stuck in the ground.

In South America, native trappers usually catch marmosets in small, self-triggered traps set up in the trees. For tribally-organized simians such as squirrel monkeys, they bait fishtrap-like baskets; escapes are prevented by the use of funnel-shaped entrances that have been spiked to allow inward traffic only. Similar traps are used in Malaya for the capture of pigtailed macaques.

There is nothing quite as cute as a baby lion.

Trappers seldom attempt to capture adults of the larger species of primates. Adults would prove highly dangerous to their captors under almost any circum-

A baby lion makes quite an armful, as Naia Haast from the Miami Serpentarium finds out. This little fellow already shows signs of immense strength.

stances, and the ordinary native trapper would be unable to construct a holding cage strong enough to keep any of the Great Apes or a baboon patriarch. The commercial value of baby gorillas and chimps, however, is well known, and native trappers rely on ancient method in capturing an infant: they kill its mother. Adults could be subdued with tranquilizer guns instead, but the typical trapper has neither the professional training nor the equipment that would be needed available; he must follow the old methods.

No matter what type of snare is used, the monkey's life is probably imperiled, unsuspectingly, by the trap itself. In lands where basketry and trap-making are ancient skills, traps are used repeatedly, without sanitizing of any kind, and infections are passed along from one monkey to the next. Some of the commonly used building materials, such as split bamboo, have sharp edges, and the cages and traps cause cuts to the monkey's feet and hands, increasing the chances of infection.

The newly trapped monkey probably spends several weeks in a native village while his captors gather more specimens for delivery to a collecting agent. During that time, the animal is exposed to whatever human diseases are prevalent and to the parasites of the local cat and dog populations. Whatever nutritional difficulties the villagers face will undoubtedly face him.

In Central Africa, baby chimps and gorillas are usually nourished at the breast of a village woman who is nursing her own baby. The common unbalanced nutrition facing the human population and the extensive parasitism from which they suffer is reflected in the potbellies of their own infants and those of the baby apes they have reared.

During the entire period of his entrapment, the monkey or ape is handicapped in fighting off the dangers which threaten him, and his physical condition may begin early to decline. Stress in monkeys delays the development of some antibodies and increases the likelihood that they will be unable to throw off infections and parasites their bodies normally could control.

The captive monkey is faced with many stressful conditions from the early moments of his imprisonment: he has been removed from his social and natural environment and placed in an unfamiliar situation. The food and water he receives are unfamiliar; he is held with unfamiliar cagemates. He faces a succession of unfamiliar surroundings. One of his reactions is the copious elimination of a "fear stool," day after day after day—a physical reaction that leads to increasing dehydration of the animal's body and places his life in jeopardy.

The monkey's health may have begun a serious decline by the time the trapper has delivered his merchandise to a collecting agent for weighing and sorting and received his pay. A new threat may face the animal as he is placed in a reused and possibly contaminated cage and shipped off to the animal exporter's compound.

Transportation

The compounds of exporters are generally well kept and reasonably clean, according to W. T. Roth, writing in *Primates in Medicine* (S. Karger, 1967), but groups of animals previously exposed to various infections may be held together in gang cages and placed with newer arrivals. Transmission of diseases can snowball in such a situation, he points out.

At one time, when animal shipments were made by lengthy boat trips, the death of a majority of the animals shipped was not uncommon. The advent of the 24-hour jet transport has considerably eased the problem of transportation deaths, but it has not wholly solved the difficulties of delivering live animals to foreign countries. It may have introduced a new hazard: the series of rapid changes of climate to which a captured simian must adjust.

Roth describes a flight he made with some 300 Green Monkeys shipped from Ethiopia by way of Beirut and Frankfurt to New York. The journey began in the Ethiopian highlands on a chilly morning, after the 300 grivets had been handcaught from their holding cages in the compound of an experienced exporter.

The grivets were placed in individual compartments in a crate which held four animals each. Otherwise, the exporter had learned, there might be severe losses from fighting between animals. The crates were carried by truck for 90 miles to the airport at Addis Ababa. As the sun climbed higher, heat built up in the closed truck, but the exporter had learned that losses were higher if the animals were subjected to drafts and dust during the trip. At the airport, export formalities took two hours. The crates were then unloaded from the truck and lined up for loading on to the airplane. No shade was available, although the temperature climbed to 105°; but air freight regulations demand that the cargo be ready for loading two hours prior to the scheduled departure of the craft. During this time, Roth saw to the feeding and watering of the animals. Eventually, the crated animals were loaded into the plane and the doors closed. There was no electrical ventilating equipment in operation because international flight regulations require that all of the power supply for planes be shut off while on the ground. Had there been a delay in leaving, Roth points out, the shipper could have requested that ground-powered ventilation equipment be set up to provide fresh air in the cargo hold. Plane and cargo departed promptly, however, landing at Port Sudan for refueling. Here Roth found the same ventilation problems, and airport officials had to find the equipment; ventilation actually began some 45 minutes after the plane touched down.

At Beirut, the crated animals were unloaded and transferred some three hours later to a jet freighter. During that time, the attendant fed and watered the grivets.

At Frankfurt, engine problems forced unloading of the cargo for transfer to another carrier. Here Roth found icy January winds, as the crates were loaded onto open cargo carts to be wheeled to the other side of the airport. The attendant secured blankets and canvas to cover the crates to protect the animals from the cold weather, but, Roth points out, these would not have been provided by the airport personnel if they had not been specifically requested.

Finally, 18 hours after the journey's start, the ship-

People who want to keep exotic pets should certainly be allowed to do so as long as the animals are kept properly. Who wouldn't want a baby lion?

ment touched down in New York—but the problems were still not over. The weary attendant still had to convince cargo department employees not to leave the crated animals sitting outside in a New York snowstorm.

Roth has proved that it is possible to deliver 300 simians with only one casualty or subsequent loss where adequate provisions are made. However, *adequate provisions* meant starting with reasonably healthy animals and *sending an attendant* with the shipment of monkeys.

Few shipments of monkeys are so fortunate. Most often, animal shipments are wholly dependent on the concern and knowledge of a series of busy airport employees.

W. T. Roth states that the casualty rate on capturing and shipping monkeys could be substantially reduced if the following recommendations were followed:

1. Increase the value of the animal from the trap stage onward, together with increased demands for the physical condition of the animal.
2. Know the biological entity of the subject so that you can properly judge the animal's condition.
3. Expect to pay a higher freight charge for improved transportation methods, including the use of qualified attendants.
4. Provide more spacious holding equipment and specially oriented nutrition to restore primates to nearly normal levels.

Because of the threat increased commercial and medical use of simians poses, he urges the establishment of more nature parks and game sanctuaries and the development of controlled breeding reservoirs for monkeys and apes [reprinted, May, 1970, the *Simian,* monthly publication of Simian Society of America, Inc., 6824 North 38 Drive, Phoenix, Arizona 85019].

Thus it can be seen that the frightened and worried little faces of monkeys offered for sale suggest the terror and heartbreak that they have gone through before you get them. Their human-looking faces almost cry out in pity. Those who really love them should get into the field and see exactly how their pet is acquired. If the only way it can be obtained is by shooting the mother, then they should be left in the jungle. The adults could be shot with the popular dart guns that subdue but do not kill the animal. Killing an animal to steal its baby is a disgrace to the animal trade. It should not be tolerated by animal lovers. It is an inhumane and wasteful way to acquire a wild animal, and animal dealers and importers should look into the methods by which their animals are trapped to eliminate these cruel collectors from their business. It has been said by investigators who have gone into the field to check on animal compounds and trappers that for every gibbon that reaches its ultimate destination alive and in good health, as many as a dozen or more are killed or die in obtaining that one specimen. For ocelots, the figure is said to run even higher. No wonder the conservationists are up in arms over the matter. It is a disastrous waste of wildlife.

I have always been interested in wild animals from the first day that I became aware of them. I always thought animal dealers should love animals. That is why they are in the business, I reasoned. But it is not so. Many animal dealers *do* love animals and would no more mistreat an animal than they would their own child. But this is not the case with all of them. A number of animal handlers are in the business only for the money. They care little for the health and happiness of their charges. It is a pity that they cannot be caged and left in the custody of their animals for a while. Perhaps then they might mend their ways. Some dealers are totally heartless. They would leave a whole shipment of animals at the airport at night to freeze, or to die of thirst and dehydration in hot weather, rather than pay overcharges on customs to clear the shipment. Instead of doing their best to save the animals, they simply file a bad-order claim and write off the animals as a business loss. They ship their animals on the coldest days of winter in a thin, cardboard box. Some of their unfortunate charges are dead before they even leave the airport. Others arrive sick and shaky at their destination. They are frightened and dehydrated so badly that no one could possibly save them except an experienced veterinarian who knows how to handle exotic animals. Consequently, many wild animals die after they have left the jungle, and conservationists howl for new legislation.

A baby lion makes a lovable pet. Keeping an exotic animal can be a wonderful experience.

But this is only one side of the picture. There are other animal dealers who take exceptional care of their wild charges. They insist that the animals be shipped to them in large, airy cages. They meet the plane when the animals arrive and hustle the animals back to their headquarters, where the frightened beasts are placed in clean, roomy cages. They feed the animals the very best of foods, and they ship them out only when weather conditions are favorable to assure safe transport and arrival. Such dealers should be commended for their efforts, but this is not the case. They are often condemned right along with the quick-profit animal hustler.

There are also the so-called roadside zoos that enter the picture. Again these are the target of "animal lovers" and conservationists who have nothing but bad words for the small roadside zoos. But all roadside zoos are not bad. Some of them treat their animals better than even the big public zoos, and this is a fact. They keep their animals in huge, natural outdoor cages that most zoos could not begin to duplicate, and they give their animals personal care, second to none. I read one scathing report in a national magazine about roadside zoos that was sickening. But I would not say that this is typical of all roadside zoos, as the article implied. I wrote to the magazine about my own views on the subject, having visited many roadside zoos and also many large public zoos. They did not print my letter though, especially since I pointed out that all roadside zoos are not bad and that I could find mistreated and ill-confined animals even in some of our large public zoos.

I abhor mistreatment of animals. I abhor even more the horrible torture inflicted upon animals by medical and scientific laboratories in the name of medical research. Some of it is necessary, but much of it is mere duplication of experiments that have been done hundreds, even thousands, of times. I often wonder why the superconservationists rarely attack the laboratories that willfully inflict painful torture upon thousands of animals each year. I know some of it is necessary, but I feel that a great deal of it could be eliminated without seriously affecting progress.

An exotic animal is a wonderful creature. It is so different from the tame, domesticated animal that has lost much of its spirit and originality. It is utterly fascinating and unpredictable. This is exactly why people want them, for they are *different* from the housecat or the pet rabbit. Although I consider myself a strict conservationist, I would not like to see any total ban on the importing of animals. I would of course like to see a ban on the killing of spotted cats, alligators, and crocodiles for their leather and pelts, but banning of all exotic animals from the public would be a sorry thing. The worst part about it is that if the animal ban is put into effect, it would keep the animals out of pet shops and private homes, but the animals still could be brought into the country by special permits for all kinds of reasons including laboratory research. Animals are electrocuted, roasted, starved, or poisoned. Some have their bones slowly broken. A little of this should go a long way! Animal behavior labs, which study the living animal without harming it, are not in the same league. Their work is commendable.

At this writing, many or all exotic pets, including monkeys, are now tentatively banned from import in Great Britain. California is trying to pass (or has passed) a law to halt the importing of all mammals by pet shops. This would include any animal not native to the state. Florida is following along with legions of new laws. Zoos would be excluded from the ruling. I am sure that zoos are excluded in England also. This is not fair! Such unfair legislation, even though introduced with humane forethought, should be rescinded. Many states are now passing laws strictly regulating or banning altogether the keeping of animals by individuals, but the loopholes in the law allow special groups to keep all the animals they want, with no questions asked. If we are to call our country the Land of the Free, we must have fair laws that do not cater to special pressure groups.

The so-called Endangered Species Law must be examined with a judicious eye to see that it is not distorted and made the tool of special groups that will be able to keep all the animals they need to the exclusion of everyone else. We must question the facts as to whether or not an animal is really endangered. One example is the orangutan. Is it really an endangered species? It lives in some of the world's most remote areas, huge forested jungles that have not even been explored yet. I have talked with scientists who visited the land of the orangutan and who found it difficult to believe that the animal could be in danger. Is the gorilla really endangered? The giant anteater and the giant armadillo—are they really in danger? They are all on danger lists, yet it is difficult to believe that animals in the vast Amazon jungles or in the remote jungles of Africa can be in danger of extinction. While it is true that their habitat must be preserved, banning the importation of them is misguided conservation. If they are too plentiful in their dwindling habitat, their numbers should be thinned out. The surplus should be sold to zoos, animal farms, or individuals, who would try to raise them to propagate the species. They should never be shot, except as a last alternative, and even then, they should only be shot by hunters rather than by game-park officials. If animals must be thinned out by culling, why not let hunters shoot them out? They could pay a high license fee that could be used to further protect the species.

I do not think it is remotely fair to allow only zoos or members of their specialized society to be the only ones who can keep such animals as the orangutan. If someone else is qualified to keep the species in good surroundings, he should be allowed to do so, if fairness is to be considered. In our "Land of the Free" we are getting so many laws that we are rapidly becoming the most tightly regulated country on earth. There must be a reverse of this trend or we soon will not be allowed to whistle when we walk down the street. One animal dealer recently told me that new, pending regulations would require that he secure a special license or permit for every single animal he sells! This is ridiculous—as if the poor businessman does not have enough to do already. The bureaucratic red tape and ever-increasing regulations have a pink tinge and do not belong in a free country. I finally gave up my own specialized marine-collecting business at which I worked for two decades because I was regulated out of business by the Florida Conservation Department, who over the years gradually ruled that everything I did was against some law or another. Yet this same "wildlife protection agency" has allowed our state to be polluted and destroyed beyond belief. The conservation agents would go around measuring bass, while the lake they were

The baby lion's playful snarl is cute when the animal is small. But remember, it will soon grow into a powerful beast.

Naia Haast frolics with her pet lion at the Miami Serpentarium.

on would be totally destroyed by pollution. Does this make any sense? Not much! We must guard against unfair, fanatic, and ill-advised legislation regarding wildlife, native or foreign. We are all entitled to share in its enjoyment.

Mr. Average Human will not be able to buy a pet monkey, ocelot, or any other exotic pet when new legislation finally goes into effect. But let us hope it will never happen. I think laws should be passed regulating the humane trapping and handling of wild animals. The shipping of said animals should also come under much closer scrutiny. Also, of course, any animal in real danger of extinction should certainly be protected in its natural habitat until its numbers have increased to a safe level. Then a few could be sent to zoos or animal farms with the proper grounds to house them so that they will reproduce in captivity. (Many zoos do not have sufficient acreage for this.)

As for a pet monkey or ocelot depleting the stock in the wild, this matter should be investigated by an unbiased group. Perhaps the trapping of monkeys has made no dent in the overall population. Many of them are purchased from Indians who consider the monkey only as food, the same as we might consider a wild rabbit or quail. It has been reported to me by reliable animal dealers that most of the monkeys sold today come from a very few areas, and that often the monkeys are caught in farming areas where they are regarded as destructive pests. Also, it is general knowledge that the tropical rain forests and jungles in which the monkeys and ocelots live are fantastically extensive, covering over two million square miles. It would appear to me that it would be most difficult to seriously affect the overall monkey and animal population by collecting a few thousand a month for pets. But the facts should be investigated by an impartial group, and if any species of animal is becoming even slightly scarce in a given area, then a closed season on that animal in that area could be enacted. Having studied wildlife conditions around the world for many years, I feel that it is not necessarily the animal collector who is causing the demise of the animals, but rather it is the land developers, the tropical cattle rancher, who totally destroy the forests and woodlands, particularly in the tropics, who slash and burn vast tracts of virgin woodland, thereby systematically wiping out the animal life in the area forever. Yet we seldom hear a cry against these get-rich land developers who drain swamps, bulldoze down miles of beautiful hardwood forests, and convert the land into housing lots. Nor is the federal government ever brought into the picture. It is a fact that our own Fish and Wildlife Service (Division of Wildlife Service) still poisons many animals each year by setting deadly cyanide-loaded shells into the ground with a piece of scented cotton to attract wolves and coyotes. Yet both the wolf and the coyote are considered endangered species. Then, as if this is not bad enough, after the service has eliminated the coyotes from an area, it notes that the field mice and other small rodents, upon which the coyotes feed, have now increased to dangerous levels. So, it then sprays the entire wild area with poisoned grain to kill the rodents. Of course many birds eat the grain and are summarily poisoned. Even endangered species of birds, such as the bald eagle, may feed on the poisoned rodents and eventually die. Consequently, we have a government agency telling us not to import animals because we may deplete them in their natural area, while this same agency poisons huge areas killing off what little wildlife we have remaining in our *own* country. Some hypocrisy! The service should erase this blemish from its fine department (see "Appendix A").

Each year, so much new legislation is passed against the keeping of animals that I truly wonder where it will end. I wonder if we are not headed for a totalitarian state where everything we do must be regulated by a special license or permit. I would like to see a reversal of this trend. Already some states have passed laws prohibiting the importation of practically all animals for pets. I feel this is unwise and unconstitutional, and those states which are affected should call for a repeal of such laws. An exotic animal is a special creature. It can bring joy and excitement to an otherwise dull life for many individuals. Of course it can bring problems too, and one of the purposes of this book is to help point out the fact that exotic pets are not all happiness. There are problems in keeping them, for exotics are as different from a cat or dog as a toy cap pistol is different from a sixteen-inch cannon.

The very word *exotic* means "of foreign origin, not native, striking or unusual in effect or appearance," according to the Random House Dictionary of English Language, College Edition. The last description, "striking or unusual," precisely describes the exotic animal, and this is why people desire them. There is hardly a more handsome animal than a bright-eyed ocelot with its shiny coat and sparkling personality. Nor is there any pet quite as lovable as a young woolly monkey. It even looks human and craves affection for its very existence. It is a marvelous animal, as are all of the other monkeys. Their almost human traits make them a coveted pet and many people find delight in keeping them. I do not think it is fair to deny the general public the right to keep one as a pet, so long as they care for it properly. But before you buy one, send fifteen cents to the Simian Society of America for its informative booklet on monkey care. It gives you the basic information on selecting and keeping your pet and is highly recommended. Nearly everyone finds the monkey irresistible. Even travelers writing for *National Geographic* bring home pet monkeys that they acquire on their trips through jungle areas. A recent article in the August, 1970, issue tells of an explorer who adopted a baby capuchin monkey while in Colombia. He just could not resist the cute little fellow who would "peek over his shoulder" as he typed his article for the famous magazine.

It would not be fair *not* to be able to keep these wonderful creatures. After all, the individual can give a monkey far more personal care than any zoo. People should learn to keep the animals properly. Education is important. No pet shop or animal dealer should ever sell an exotic pet to an individual without first telling his customer precisely what he is in for when he purchases the animal. It should be explained to him that if the animal is small, it will soon grow, and that when it is fully mature it could prove difficult to handle. Anyone who buys a baby lion or a chimp should do so with the thought in mind that it is strictly a temporary situation, and that a parting of the ways is almost inevitable when the animal matures. The average person can easily handle a baby lion or chimp, but few people can manage a full-grown lion or chimp when it gets out of hand. You will have to either sell your prized pet to an animal dealer or donate it to a zoo. And very often the zoo will not even want your animal for free.

Today you can buy a book about practically any animal at your local pet shop. Many of the smaller books cost a dollar, but some may cost several dollars. A prospective owner of an exotic pet would be very wise if he invested in a book on the animal he wishes to keep *before* he makes his purchase. It may open his eyes to the fact that his pet may not be all joy and will require special foods and clean drinking water daily for seven days a week, 365 days a year. This alone may deter him from acquiring a strange pet, and he may decide that it is easier to visit the zoo. I shall never forget the day that I almost bought a pair of mandrills for my backyard. I was visiting Bill Chase, a local animal dealer of high repute, when I noticed the cute little animals in one of his cages. They were adorably tame and I wanted them badly. But Bill advised against it, for he knew that the animals would grow into powerful beasts that certainly would not belong in anyone's backyard. A full-grown mandrill is as dangerous and as treacherous as any animal in the wild. I have always been thankful that I did not bring that pair of animals to my home.

I have kept all types of animals in my lifetime, including skunks, woodchucks, squirrels, and weasels, as well as many exotic species. At one time I worked at Benson's Wild Animal Farm in Hudson, New Hampshire, and almost became a lion trainer under the G.I. educational bill. I have also owned several pet shops and have kept my share of exotic pets. Therefore I am familiar with the problems of keeping them, both in the confinement of a cage and in larger outdoor enclosures. I have a special feeling for animals that is no doubt shared with many of the conservationists, and I sometimes feel that they would all be better off if they were left in the jungle. But there are other things to consider, such as the fact that very often many of the animals sold as pets or kept in people's home zoos would probably die prematurely even if left in their own country. The forests are burned, many animals are shot as pests, and others are poisoned by aerial sprays that kill and defoliate the tropical forests so that the land can be used for agriculture or housing.

Can you imagine how many animals have been killed in Vietnam by the fantastic destruction of that nation's forests by our own government, with its aerial sprays, napalm, and bombs? This is one of the few areas in the world where a rare rhinoceros still lives—if any have survived the holocaust. Therefore, perhaps many of the imported animals would not have survived anyhow, had they been left in their native land. Do not forget that monkeys are the chief food of many natives, as well as of the jaguar and other carnivores. What we should do is strive for a higher survival rate of all animals brought into the country. People should be made aware of all the problems that they will face when they acquire a wild pet. They should have a designated minimum-sized cage for each species. I think a small two-by-four-foot wire cage is entirely inadequate for an ocelot, monkey, or most any wild creature except a rodent. Six-by-eight feet is more appropriate, with trees for swinging and jumping. Ten-by-twenty feet would be even better as a cage for a wild animal so that it will not develop cage paralysis as it might in a small wire cage. A spare bedroom with a two- or three-inch floor covering of peat moss or sugarcane pulp and several good-sized tree limbs or green potted plants would make a nice home for an exotic pet. But the most important thing to remember is to buy a book about the animal that you intend to purchase. Read the book thoroughly, and then you will be able to decide whether or not you will want to acquire the animal.

Perhaps your local dealer will rent you a pet for a specific period of time so that you can see if you want to keep it or not. This is a superb idea, and many pet shops do rent animals. This way you can try out the animal of your choice to see how it fits into your family life. Also, if you are really gung ho about animals, perhaps you could get your local neighborhood to build a minizoo. You could purchase a vacant lot, overgrown with trees and brush, and convert it into a small zoo for all the people in your area. The average-sized city could have hundreds of minizoos, and they would be very inexpensive to maintain compared with the larger zoos, for all the feeding and caring of the animals could be done on a voluntary basis. It would benefit the whole

Tony Gomez pets a friendly tapir. Tony is an animal dealer who truly loves animals, as does his associate Bill Chase, both of Miami. They import everything from anteaters to elephants for nation's zoos.

community, and if it were financed by a thousand individual homeowners, the cost per house would be practically nothing on a yearly basis, hired help included.

Big zoos are fine, but invariably they are so far away from much of the residential area of a city that only those living close to them can really enjoy them on a regular basis. Here in Miami we have the Crandon Park Zoo. It is a pretty zoo, but it is located out on an island. It is about thirty miles from our house. To get there on a Sunday is practically impossible, for there is a traffic jam of cars that backs up for ten or fifteen miles at times, making a three- or four-hour wait just to get to the zoo! I would love to have a minizoo in our neighborhood. I am certain that many of my neighbors would too. Almost everyone likes animals, but with today's traffic, we need zoos closer to home so that people can get to them easily and visit them more often. The minizoo might just be your answer if you love animals but do not want to take them into your home. If a number of homeowners get together and purchase some vacant lots or even a good-sized piece of land if they live in a rural area, they could easily make a fine minizoo that would be the pride of their community. They could keep all kinds of exotic creatures there, with the exception of the

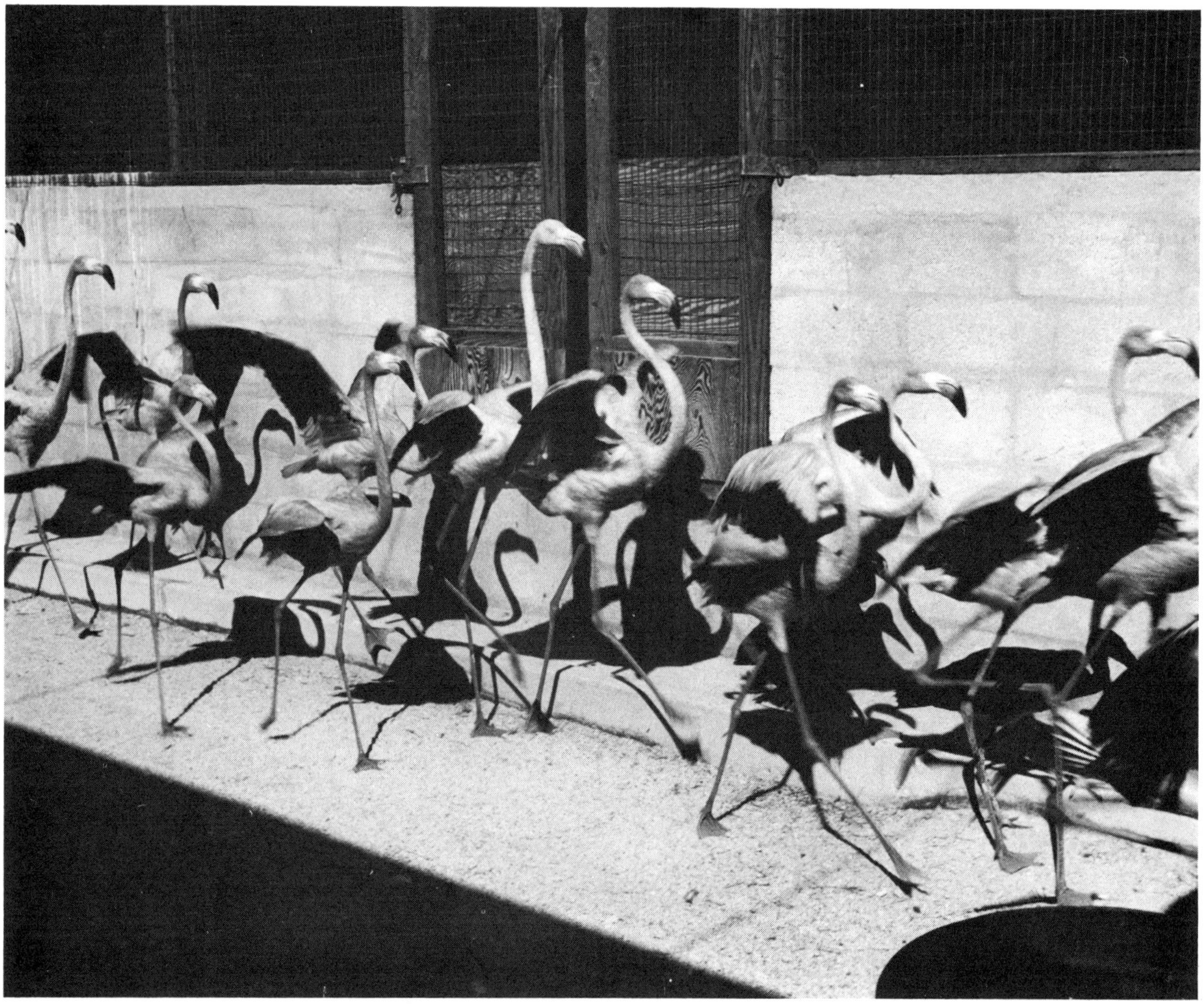

A flock of red flamingos at Chase and Gomez's animal compound in Miami. Note that the birds have clean, spacious quarters and are in excellent health.

dangerous cats and elephants. These could be left to the big zoos, since they are better equipped to handle them. But they *could* keep a small elephant for at least a few years until it outgrows the area. This of course would depend upon the size of your local minizoo.

You would of course have to work out the legal aspects for such a project, but they should not be insurmountable. You would also need some skilled animal workers. You may get a little friction from your local zoo, who may view you as competition. Unfortunately, some zoos take the attitude that no one should keep animals except them and that no one else could possibly take care of animals as skillfully as they can. But on the other hand, your local zoo may welcome your project and even assist you or work with you. The goal of both your zoo and theirs is to keep and display animals for the public, in a natural, humane way. Also, there is the breeding of rarer species to be considered, and minizoos could contribute a great deal in this area. They could specialize in certain species and, if successful in breeding them, could offer their surplus to other zoos throughout the land. Thus they would be aiding conservation in an area where it is urgently needed. At the same time, your exotic pet could be kept there. There is room for many more zoos than we have today.

But whether you buy your pet, rent it, or keep it in a minizoo, you should strive to make the pet feel at home. Provide it with as large and natural a cage as possible and take care of it to the best of your ability. Learn as much about its feeding and living requirements as you can. If you like exotic cats, ocelots, and similar species, remember that all male cats spray urine in their surroundings as a natural part of their life to mark off certain areas as "off limits" to other male cats. A large cat like an ocelot will make your home smell very unpleasant. Also, keep in mind that cuddly animals like the coatimundi, kinkajou, and similar animals are nocturnal. They sleep a great deal during the day but at night they are wide awake and full of playful (or serious) destruction. They may tear your house apart when they go on a rampage. If you keep them in a spare room, the walls should be covered with tough wallboard so that they cannot tear it up with their strong claws in their search for insects.

If you buy a chimp, do not forget that he will soon grow up and become stronger than you are. The same is true of a baboon or a mandrill. When they become full grown, they can be extremely dangerous. A baby elephant is cute, but just by his weight alone he can be dangerous. He could easily fall through the floor of a house, should he follow you inside. Make certain that there are no electric wires, ceiling lamps hanging down, or exposed wiring in the area in which your animal is kept. Should he chew on these he could be killed or start a fire. Elephants in particular will pull on the wires with their trunk, and with it they can reach unbelievable heights.

I personally feel that the exotic pet can be a wonderful experience in your life. They are utterly fascinating. But you must learn to cope with problems that they may present. It is a fact that they will not be all joy. The precise difference between an exotic animal and a domestic one is that the exotic animal may be constantly trying to get away. Should it escape, it will mean trouble for you—and very often, will mean death or injury to the animal, either from local dogs, automobiles, electric wires, or local law enforcement agencies. There was a case where a beautiful cheetah jumped off of a man's boat here in Miami, swam to shore, and immediately began running after people. He finally chased after a policeman on a motorcycle and was shot. The animal was totally tame and was a pet, but the people whom he chased did not know that. It was a sad thing that he had to be killed, especially since this animal is on the endangered-species list. Many monkeys and chimps are shot after they have escaped, for although they may be easily handled by their owner, when they get out they often are hostile to those who may attempt to capture them.

In the ensuing chapters, I will tell you of my own personal experiences with a number of exotic animals from faraway lands. Perhaps this will be of benefit to you in deciding whether or not you should acquire an exotic pet. I found the strange creatures intensely interesting. It was always a pleasure to see them, but they presented me with problems and a tear or two, as you will see. One thing is certain: keeping them was *not* all joy.

One last thought in defense of the exotic pet is the fact that the sincere owner of an exotic pet, particularly the child, is very likely to become the zoo director of tomorrow. Only by contact with the exotic animals of the world can man really get to know them and appreciate all the more the need for not only conserving them in their natural habitats, but equally important, the need for breeding them in captivity by pet owners, animal exhibits, and zoos.

Appendix A

Predator Control: A Study in Overkill

By Michael Frome, Conservation Editor, *Field and Stream* magazine

(The Director of the Bureau of Sport Fisheries and Wildlife, John Gottschalk, recently resigned his position in order to join the new National Marine Fisheries Service. The following becomes especially timely in shaping future wildlife policy at the Department of the Interior.—MF)

It's hardly necessary to remind readers of Field & Stream of the deep personal meaning of a dog to man or child. Our own dog, Blackie, died in August. He was an old boy of no achievement whatsoever and limited talents at best, but he added a special quality to family life. I miss him particularly now as Christmas nears. However, he was ailing pretty badly with the afflictions of advanced years and it seemed the right thing to have him put to sleep.

The story is quite different, I regret to say, concerning the loss of a prized 3½-year-old hound owned by B. B. Fant, a rancher who for over twenty-five years has been raising shorthorn cattle, sheep, goats, and horses in west Bell County, near Salada, Texas. His dog was done in by a nasty Federal bureaucracy hiding behind the high-sounding, thoroughly misleading title of Division of Wildlife Services and behind the shelter conveniently provided for it by the leadership of the Bureau of Sport Fisheries and Wildlife and of the Department of the Interior, of which it is a part.

The Division of Wildlife Services has one prime goal at the root of its existence: to kill wildlife. It gets away with murder, particularly of the nation's rich heritage of predatory animals—wolves, mountain lions, coyotes, bobcats, foxes, badgers—as well as anything else that may be handy, such as Mr. Fant's Walker hound. It conducts its dirty work free of any serious criticism from the national wildlife conservation establishment. Only a spunky outfit called Defenders of Wildlife has shown the guts to lay the issues on the line and to press for a major overhaul in behalf of the public interest.

"Man today possesses tremendous capability to alter and destroy natural habitat, yet we seem to be ignoring this fact in the education of wildlife professionals and in the conduct of the programs of which they ultimately become a part," Daniel A. Poole, president of the Wildlife Management Institute, declared recently in a timely soul-searching exercise. "Wildlifers appear either reluctant or ill-equipped to take on the forces that really have an impact on the resource."

How right you are, Daniel. The continuing environmental deterioration must be countered with a major crusade for wildlife, based upon the direct relationship of the state of wildlife and its habitat with the health of society. The crusade should rightfully be led by the Bureau of Sport Fisheries and Wildlife, considering the Secretary of the Interior had stated repeatedly that government's role is to lead. And there should be strong support, with pressure where need be, from the wildlife conservation establishment. It just isn't happening that way, however. The comfortable approach is to conserve fish and game by press release, avoiding confrontation with manufacturers of chemical poisons, the wood products and pulp industry, and the other forces that really have an impact on the resource, including the self-perpetuating bureaucracy itself.

Mr. Fant has owned and trained stock dogs and hunting dogs all his life. The particular hound in question he had raised from a pup. She was "stock broke" and an all-round hunting dog. One day a few months ago he was out in a pickup truck up the north line, with the hound trailing behind. After checking his cows and calves, the rancher went into his house. The dog followed. One hour later he found the dog dead and stiff at the back door. Inquiry of a neighbor revealed that a Federal poison spreader, sometimes called Government Trapper, had been setting out strychnine only two days before. Mr. Fant telephoned the Federal agent, who admitted placing fifty strychnine baits around his place, even within 10 feet of the fence.

Deeply concerned, Mr. Fant called offices of the Bureau of Sport Fisheries and Wildlife at Fort Worth and San Antonio and requested a full investigation. He felt the response at Fort Worth was impolite and inconsiderate. However, at the San Antonio office, Milton Caroline, the State Supervisor of Predator and Rodent Control, indicated that he would check the incident and write to Mr. Fant. After three months, no word has been received.

"This trapper is putting poison baits all over the country and you should see the buzzards," writes Mr. Fant to Field & Stream. "I believe the baits are killing lots of wildlife and have not killed a wolf yet. I am afraid the strychnine and what it kills will be on the ground a long time."

None of this is surprising, alas, to those who follow the trail of death left by the Division of Wildlife Services. Readers may recall my article on "Predators, Prejudice and Politics" in December 1967, and my column, "A Poison in Tennessee," in January 1970. In preparation of the latter, I addressed a letter on August 25, 1969, to John Gottschalk, Director of the Bureau of Sport Fisheries and Wildlife, requesting his verification of statements attributed to his subordinates. I am still waiting for a response, even though I sought one again recently through the Office of Information of the Interior Department. This brushoff is part of the ingenious system by which public servants treat the public as poor relations who have no right to know what's going on.

Horned screamers and a pelican at Chase and Gomez's compound. Sometimes there will be nearly as many wild animals (including lions and tigers) at the compound as there are in an average zoo.

The attitude of the leadership of the Bureau of Sport Fisheries and Wildlife toward the special economic interests is another matter. The Subcommittee on Conservation and Natural Resources of the House of Representatives has been conducting an investigation of the Bureau's role in the wildlife kill in Tennessee which I had described. In a long letter of justification to Representative Henry S. Reuss, the Subcommittee chairman, Mr. Gottschalk ended with a paean of praise to the Bowaters Corporation, whose practices I had criticized. It represents, he wrote, "an outstanding example of both a private industry's cooperation with conservation agencies and the continuing interest of a private industry in high quality recreation." This about a firm that has eliminated wildlife habitat, indiscriminately spread hazardous chemical poisons, and is now being sued by a citizens group for dumping pollutants into the Hiwassee River.

If there is one field in which the Interior Department should truly be outstanding and inspirational to all Americans, that field is wildlife. The effort to protect rare and endangered species, which should be the showpiece of commitment, is rather dismal. The predator control programs in the West are far worse. Agents lack scientific training or environmental skills; they do not conceive of their work in terms of animals saved or crops saved, but in the number of predators killed—as though this marks some special achievement.

Director Gottschalk and his superior, Assistant Secretary Leslie Glasgow, have assured me on occasion of their desire to halt the widespread indiscriminate killing conducted under the heading of predator control. They say they'd like to cut the program, but that the woolgrowing industry prevents them from doing so through political influence in Congress. Therefore they tolerate, sanction, and continue to sponsor it.

The people are far ahead of the ingrown Washington bureaucracy, as I have stressed in this space before. An article by Larry Roop in the September 1970 issue of Wyoming Wildlife thoughtfully explains the role of the predator and the wasteful nature of control programs. "No one who lives in cattle and sheep country can deny that some limited control is necessary where predation on livestock occurs," he writes, "but all too often there is such an overreaction that expenditures on control far exceed the value of lost sheep or cattle. Nor can the small amount of predation justify the dropping of poison from airplanes in the wild areas of our National Forests—poisons that kill everything that eats meat, including marten, wolverine, badger, foxes, Canada jays, and many more." If this type of constructive material can appear in the publication of a state fish and game department, why should the Federal bureau remain chronically silent and supine?

An assorted collection of large birds, including rheas and storks, at Chase and Gomez's animal compound in Miami. The cages are clean with plenty of room for the birds to move around while awaiting shipment. It is the finest, best-designed animal compound that the author has seen, and a tribute to Chase and Gomez, both with a lifetime of animal handling experience.

"Predator Control in Nevada—a Pork Barrel Threat to Environment" is the title of a meaty little leaflet written and published by Dr. Robert V. Broadbent, member of the Nevada Fish and Game Commission. "I'm not arguing that the livestock people shouldn't be able to protect what is theirs," he writes. "If a band of renegade coyotes is threatening a lamb crop, then the sheepman should be allowed to take remedial action. But with his own money. It's not fair for the rancher to expect the general public or sportsman to pick up the tab. After all, tax dollars aren't made available to other private businesses to increase their margins of profit."

Dr. Broadbent makes the important point that three cents out of every dollar which Nevada sportsmen pay in license fees are used as matching funds for the cause of wildlife eradication. That money should be used for habitat protection, where it is needed desperately. Equally shocking, this is the sole fish and game expense that completely escapes accurate cost-accounting—the funds are simply delivered in a lump sum to the Division of Wildlife Services, which runs its show on the basis of money available instead of proven need.

But the natives at the grass roots are restless. This nefarious system is clearly nearing its end. The Nevada legislature in January will consider proposals to wipe its slate clean and to start afresh in considering the wildlife interests of all the people. Conservation groups are strongly backing the new look; in fact, the National Wildlife Federation will do well to follow the lead of its Nevada affiliate. (Incidentally, Dr. Broadbent has made available a supply of his leaflets. Write to me at FIELD & STREAM for a copy.)

What we need everywhere is the desire and passion to conserve the life-forms, all of them. The widespread, indiscriminate poison-killing of one type of animal to protect another makes no sense. There is a vast difference between slaughter and selective control. This is something the sportsman has always known—a distinction he has contributed considerable money and effort to maintain. Shouldn't the Government with its vast resources be expected to do the same?

A father teaches his son about reptiles. Warren Prince, reptile expert, drapes a huge snake over his son, Warren, Jr., who is growing up to become a naturalist just like his father. If all exotic creatures were outlawed, people would have no contact with them and would lose interest in their welfare. Courtesy of Warren Prince.

Appendix B

SHOULD YOU HANDLE EXOTIC PETS?*
By Donald R. Collins, D.V.M.

Six step program explains what you must know before selling these animals

A small child stood before me, great tears welling in his eyes and staining his cheeks.

"Please, Doctor, Freddie's sick. Please, can't you do something for him?" Gently he opened his hands and there, tenderly wrapped in his mother's newest washcloth lay Freddie, a thirteen-lined ground squirrel, hopelessly mauled by the family's cat. No one could do anything for Freddie anymore.

Poor Freddie never had a chance. Purchased from a pet shop owner who had the "chipmunk" on special, by a father who thought he was just "too cute" to pass up, Freddie had escaped almost immediately from his makeshift cardboard cage. Neither the pet shop nor the local feed store had a cage suitable for a "chipmunk." After escaping, he had run around loose in the bedroom several days because no one knew how to handle him. Finally, weakened by improper nutrition because no one knew what to feed him, Freddie was easily cornered by the cat in the strange new world of the boy's bedroom.

No one wins when a pet meets such an untimely end. All that remains is a heartbroken child, an upset family, a veterinarian with a disenchanted client, and an angry customer. Moreover, every such event adds one more round of ammunition to the growing army of people who advocate that wild animals are unfit as pets and should be left to the wild.

While I agree that innocent wild creatures shouldn't

* Reprinted from *Pet Shop Management,* July, 1969.

suffer the same fate as poor Freddie, I can't go along with the idea that denying a youngster the rewards and pleasures of keeping a wild pet is the way to solve all the problems of vanishing species, inhumane treatment, and decimation of wildlife. Quite the contrary!

Every youngster, sooner or later, makes a pet out of some wild creature. It may be as simple as putting a grasshopper into a peanut butter jar for a few hours or as complex as rearing a two-week-old fox squirrel. The urge to make a wild animal into a pet is as natural as growing up itself. I've always felt, as a wildlife biologist as well as a veterinarian, that everyone who derives his livelihood from wild animals has a responsibility not to discourage this normal urge, but instead to direct it, so that it helped develop a love and respect for all life—wild or domestic, animal or human. The *boy* that has reared a family of fox squirrels is not likely to grow up to be the *man* who carelessly throws a lighted match into the forest.

The wild animal pet should be more than an introduction to the care and feeding of livestock. It should be a do-it-yourself course in natural history. The wild pet should become an interesting and educational playmate that creates a sense of responsibility, kindness and consideration far beyond that which comes from a puppy or kitten. Most important of all, by making it my business to encourage prospective new wild pet owners to proceed in a manner that insured the health and welfare of the animal, I have seen wild pets not only make unusual and instructive pets, but become the instrument through which many people learn to love wild animals rather than to exploit them.

An increasing number of people are saying that pet shops are in the business of exploiting wild animals. They argue that the only thing a retailer cares about his wild livestock is how much money exchanges hands for it. This, they contend, is reason enough why a wild pet should never be a pet shop owner's business. I suggest that just the opposite is true, that what every pet retailer needs is to be *more concerned* with the wild pet business. I suggest that his business with wild animal pets should begin the day they are born in the wild, and should not end until the day that pet dies.

Some of the people whose business it is to conserve wildlife are saying thousands of individuals of some species die at the expense of getting one to the pet shop. I wonder about the accuracy of that remark. A pet shop owner should *know* if it is exaggeration or fact. It's his business to know the facts about wild animal importation. If he's unwilling to dig out and accept these facts, whether they are bad or good, maybe he really doesn't belong in the wild pet business.

Another group has charged that most Americans keeping wild pets obtained them as status symbols, don't have the emotional stability to cope with such pets, and are totally ignorant of their pet's physical and emotional needs. It has been my experience that most wild pet owners sincerely want to care for their new pet the correct way. They *will* take the time and effort to prepare for their pet, if they only know how. When informed of the obstacles that may appear during pet ownership they often decide to choose another pet more suitable to their needs. The problem is, most people don't get any of this information until it is too late.

During 10 years of drying tears and burying Freddies I have found the only way to avoid these distressing situations is to educate factually and prepare emotionally all prospective wild pet owners *before* they become actual wild pet owners. I use a simple six-step method to insure that prospective owners are prepared for their new pet.

Step 1. Every animal species requires its own type of housing, food and care. Until a prospective owner has acquainted himself with all three for the pet he has chosen he has no business with that pet. And if he discovers that he is unable to obtain information on the pet, he has no business buying it and should choose another pet. (This goes for retailers too, who should always gain a complete familiarity with a species before they add it to their livestock inventory.)

There are a gillion other things to be considered when choosing the right wild animal pet. Not the least of these is the prospective owner's ability to pay the price of the pet he chooses. One frequently overlooked consideration is the availability of necessary funds for maintaining proper care and feeding after the initial expense of purchase. And, don't forget the free time needed to clean the cage or pen, carry the food and water, and the wastes and bedding from the animal's quarters.

Often the most important consideration with young pet owners is parental encouragement. Some mothers who will accept a bowl of goldfish without complaint might be much less eager about a prairie rattlesnake in the TV room. The pet shop owner who allows him to purchase the rattler often finds he's exchanged a rattlesnake for a tiger by the tail. You simply can't count on prospective owners having made all these considerations beforehand. You'd better make it your business to see that a would-be animal buyer is confronted with the facts about his chosen pet and has the opportunity to consider each of them in light of his own situation. For you to be able to do this, you must be aware of all the considerations to be made when choosing a wild animal pet. But after all, that's your business, isn't it?

Step 2. The most essential step to owning a pet, wild or other, is its day to day feeding. The entire process of successfully owning a wild pet depends upon it. No prospective pet owner should even consider obtaining the animal until he has a knowledge of everything his chosen pet can, and cannot, eat. Even knowing this, unless he can be sure that there is at least a year's food supply available, he has no business with that pet. Unfortunately, if you want something done right you'd better do it yourself. This means that for every wild animal you stock you should have a year-round food supply to provide the new pet owner to feed his pet. If you do not know what a year-round food supply is for that species, or you cannot

Tarzan, in Bob Austin's backyard. The chimp had a huge tree in which to climb and play but broke away occasionally by slipping its collar. Chimps make nice pets when they are small, but they grow rapidly, and in five years become more powerful than a man. They are better off in a zoo or wildlife attraction, where they should be displayed only on water-surrounded islands instead of in steel and concrete cages.

obtain it, then you have no business dealing with that pet.

Step 3. Properly housing the wild pet is almost as important as feeding. As far as both pet owner and retailer, the things that hold true in "step two" regarding food also hold true regarding housing . . . know all about it and have it before you start dealing in that pet. There is one exception. The pet retailer need not handle proper housing for *every* wild pet he sells. As long as he knows what that pet needs, and how to build it, and he can instruct a new owner, he knows his business.

Don't overlook, either on your shelves or in your new pet owner's preparations, that every cage or pen must have the proper furnishings. Every facility for keeping a wild pet must have feeding, watering, and bedding areas. Accessory furnishings such as plants, resting materials, perches, and occupational devices can be obtained either before or after the animal. One accessory which should at least accompany the new pet home is a small lock for its cage. This will keep little brothers and sisters and other unwanted guests out. It also helps relieve the minds of the neighbors.

Step 4. Not until housing is waiting and some assurance of a yearly food supply exists, should a prospective wild pet owner begin thinking about actually obtaining the pet. When he offers a wild pet to a new owner, a retailer should know as much about that pet as any new owner that has just taken steps "one through three." In fact, the pet shop owner should have taken those steps long before he added that particular pet to his livestock list. Just as with the pet owner, the pet shop owner who has not bothered to take these steps has no business selling that species. The retailer who does bother is the one who keeps other pet retailers in the wild pet business.

Step 5. The cardinal rule in taming any wild creature is patience, both with one's self and with the animal. The new pet must be taught that no harm will come to it. Every pet should be attended at least once each day. This is done not only to feed, water, and clean, but to remind the wild pet with regular frequency that it has nothing to fear, that its owner will never do anything harmful while attending it.

Training an animal is something else again. Training is not a necessary part of step five. Some animals can never be trained. With those that can, however, doing so will keep them a lot healthier and happier.

Step 6. Successfully keeping a pet doesn't stop after it's bought, caged and tamed. A pet, wild or other, must be constantly cared for. Along with regular feeding goes regular watering, often forgotten by the novice pet owner. Something many people whose business is wild animals are not even aware of is a wild animal's need for occupational activity. The jobs of finding food, escaping from enemies, finding shelter and following water sources are denied the captive animal. This activity must be replaced by the proper recreation.

A final element is medical care. Some wild pets should have vaccinations. Parasites, both internal and external, should be watched for and dealt with quickly. Injuries and diseases should be handled by a veterinarian. Every pet shop owner who sells wild animal pets should know a veterinarian who's in the wild pet business, the treating end, that is, and should make the veterinarian's name and address available to all new wild pet owners. This will save the new owner a lot of frustrating phone calls, and will let the new owner know that you know the veterinarians in your locality who can take care of the medical needs of his new pet.

There is another group who have recently begun to make wild pets their business: the Congress of The United States. If you've been taking their activities lightly, let me suggest to you a revealing fact. Many of the same names who appeared on the recent wild life bills in both houses are the same names who a couple of years ago turned up on bills to deter dognapping and inhumane treatment of laboratory animals. Before they were through, Congress received more legislative mail on that one issue than all the other bills combined. Today their proposals are law. If you want the wild pet business to continue, you'd better get into the wild pet business with all fours . . . better still with sixes, the six steps to successfully owning, and selling, a wild animal pet.

2
Plan Ahead for Wild Pets

Do not buy a wild animal on impulse. You are certain to regret it if you do. Nearly any kind of wild animal is cute when it is a baby, but unfortunately it does not stay a baby very long. You may think a baby elephant is adorable, a baby lion or gorilla just too cute for words, and you are one hundred percent correct, for all babies are cute, including human babies. But think ahead when you are drawn to a baby lion, a bear cub, or a small baboon. Picture the adorable little fellow just two years later when it will become a powerful beast capable of inflicting serious injury to you or some member of your family. It is a fact that the little animal rolling on the floor of your local pet shop or animal farm will mature very fast. It will only be small and cute for about six months to a year at the very most. After that it will be in the "teenager" category and will start to present problems for you. Then just a few years later it will be a full-grown, powerful beast, and you may not be able to handle it at all. Certainly you can probably handle a baby elephant for a couple of years. But do not forget that you are going to become mighty attached to it, and it will return your affection in kind. However, unless you live out in the country and have a huge tract of land, you just will not be able to keep your elephant for very long. After all, you just cannot keep an elephant of any size in the middle of the city. Too many things can frighten it, and if it runs away in panic, especially when you are not there to calm it, it will run through all the fences in the neighborhood and eventually run into the street where it could be hit by a truck. An elephant will not be full grown for quite a few years, but even when it is just a few years old, it will have a mind of its own. If it does not want to move, you will not be able to budge it, even with the family car.

As for the baby lion, tiger, or other cat growing up with the children, forget it! There *are* people who keep full-grown lions around the house, but I would not want to be one of them. There are few wild animals that can be trusted completely once they mature. The cases of lions and other large cats suddenly turning on their master are legion. A veteran wild-animal dealer, Harold Schultz, from South Africa once told me about a friend of his who had adopted two baby leopards and reared them to full size. They were allowed complete freedom of the house and were fed nothing but cooked meat in an attempt to keep them from becoming wild. The animal dealer had warned the man that he should not trust the leopards, for they would some day turn on him. But his friend felt that they were completely trustworthy. One day, the inevitable happened. The leopard owner had gone hunting and walked into his house with the smell of fresh blood still on his clothes. Immediately, both of his pets attacked him and tore him to shreds. Fortunately, one of his servants was with him and promptly shot both leopards, but the man was crippled for life, a sad reward for trusting the beasts.

The same is true of other animals. A zoo keeper once told me about a woman who fed grapes almost daily to a gorilla, only to have the gorilla grab her one day and bite her severely. There are numerous cases where "tame" lions have attacked either their owner or some other person for no apparent reason. One recently bit a model while posing for a TV commercial. Tame raccoons, reared from babies, have suddenly attacked their owners, biting them severely. Even

Bob Austin kept several beautiful African crowned cranes in his fenced acre of land in Miami. The spectacular birds would fly over his house and land again in the yard to feed.

domesticated animals like cats and dogs may sometimes turn on their master or attack a neighbor.

This does not mean that wild animals will always bite or harm someone, but it does indicate that statistics for a wild animal harming its owner run rather high. There are people who have kept a lion or a puma without incident, but I would say that this is the exception rather than the rule. A baby lion is about as cute an animal as one could find, and I would not blame anyone for wanting to keep one when it is young. But as it matures, it will quickly outgrow most households (and most food budgets) and should be placed in a zoo or exhibit for its own comfort and safety as well as for the comfort and safety of others.

If you have a large, tropical backyard such as many people have in South Florida, you can keep many pets that would be difficult to keep up north, because you can keep them outside. This is healthier for them, especially for monkeys and birds, and at the same time, cage cleaning is greatly reduced. A friend of mine, Bob Austin, kept a small zoo in his backyard, trying to outdo me. He had a chimp, siamese cats, trumpeters, and several beautiful African crowned cranes. These were exotic birds, and not only did they have

Bob Austin's huge cranes would run about the yard and then take to the air. One day they flew away and never came back.

freedom of the backyard, they even flew over his house, honking loudly as they flew by. One day, they all flew away and never came back! Bob, who was a former pet shop owner, finally parted with his chimp. He told me that he had planned on getting rid of it when it was a few years old when he had first purchased it. He sold it to someone from South America so that could not visit it, since he had become very attached to it. This was a wise decision.

People who acquire baby chimps or lions as pets should look far ahead *before* they make the purchase. Buying one of these animals is sort of like having a love affair with someone, knowing that no matter how much you love her, in a few years you will have to part with her, or at least it should be this way. Chimp owners especially fall into this category, for chimps are almost human when young, and a person who raises one sort of adopts it like one of the family. But the chimp will grow up in four or five years and can become dangerous to his owner as well as to others. A five-year-old chimp is an extremely powerful creature. He may be very tame and affectionate, but he also has natural feelings and desires that may become a problem when he matures.

It is nearly impossible to confine a chimp in the

home, for it can work at chains or cages and free itself with amazing regularity. Once it escapes into a crowded neighborhood, anything is likely to happen. It may be attacked by dogs or confronted by children. The police may be called, and if the animal makes any unfriendly gesture it could well be shot. A large chimp is nothing to reckon with, and although the owner can easily handle it, should it escape away from its yard there is certain to be complications. A child may be bitten, a neighbor's property destroyed. This could lead to a parting of the chimp from its owner either willingly or by lawful means. It can be a painful experience for both chimp and owner. The chimp may be ordered destroyed or given to the local zoo, which may lead to more frustration, for the owner will find that his beloved pet is not happy and the zoo cannot or will not cater to his desires to give the animal the same personal treatment that he gave it at home.

The same is true of lions or elephants. Baby lions are inexpensive because they breed well in captivity and many people are tempted to get one as a pet. But they grow up fast, and once they are full grown they can be extremely dangerous, particularly if they escape. Ultimately they will either be shot and the owner fined or required to dispose of his oversized pet. Unless one lives out in the country and can construct a large, strong outdoor cage that will safely confine a full-grown lion, then a smaller, more suitable pet should be selected.

Since elephants cost several thousand dollars, the number of people keeping them is relatively small, since most people would not spend that much money for a pet. Also, few people in the city would even consider keeping one, if for no other reason than sanitation problems. You should live in the country or have a very large yard for grazing and exercise area if you are considering the purchase of an elephant. It is very expensive just to feed an elephant because they eat huge quantities of food, amounting to several hundred pounds a day. This will amount to a tidy sum in a year's time. Elephants are among the most intelligent of all animals. They are also the strongest. Keeping one will present numerous problems and, like the lion, a day of parting between owner and pet will occur unless the owner has enormous funds and can permanently house his pet on a ranch out in the country.

You should not keep any wild animal without learning as much as possible about it beforehand. Then you will know how to cope with it when it matures, or at least be aware of the problems facing you. Most animals grow fast, so the problem of keeping an adult of the species is with you almost before you realize it. This is also true of birds, particularly large birds like the ostrich and the rhea. Do not get one unless you have a large area out in the country where it can run to exercise. A young ostrich is one thing, but a six-to-seven-foot-tall bird weighing several hundred pounds and capable of kicking you clear out of your yard is a bird of another color. In addition to its enormous size, the adult bird may make noises that will be loud and strange, to say the least. An ostrich makes a booming noise that can be heard miles away. You may consider the noise exotic, but your neighbor may find it obnoxious and file a complaint. He may force you to get rid of your pet. People have been forced to get rid

Bob Austin with his pet chimp, which he raised from a baby, just a few months old. As the chimp matured, it became powerful and was eventually sold. Bob realized that this would have to happen before he even bought the chimp, so the parting was not so tragic.

The strange trumpeter from South America is a small, chicken-sized bird. It is totally unafraid of people and usually is very tame. It emits a low, trumpetlike noise when approached by a human friend. This beauty is Bob Austin's.

of guinea fowl because of their noise, which is only a peep compared to the loud boom of a full-grown ostrich.

If you intend to keep any large, exotic animal, it is always best to move far out into the country in agricultural land where zoning is nonexistent or at least tolerable for keeping animals. Plan ahead for your prospective pet and build as large a cage or enclosure as possible, keeping in mind that the animal will mature rapidly. Build the cage for fully adult specimens even though you may be starting out with baby animals or birds. Make certain that the enclosure will be strong enough to restrain the full-grown animals or birds even when under stress, should they charge the fence in a moment of fear or panic. A look at the cage or enclosure at your local zoo, for the kind of animal you intend to keep, will give you a good indication of the structure you will need. You can also ask the zoo keeper for his advice as to what you will need for your own specific purpose. Sometimes animals are kept in cages at the zoo not because it is best but rather because it is all they have. The zoo keeper might prefer something else if he had an opportunity to choose. Also be prepared for him to discourage your keeping of the pet altogether, as this seems to be a trend among zoo keepers.

Wild animals or pets must also be given consideration for the seasonal changes of the weather, particularly in northern climates where the winter is severe. You may acquire your animal in the warm summer months and it will appear happy and contented in its enclosure, but when the harsh winter arrives with its freezing weather and deep snow, your compound will take on the aspects of an Arctic igloo, hardly the place to keep any kind of tropical animal. You must have warm, dry winter quarters for your pet unless, of course, you intend to keep it in your house. Large animals should have a good-sized, heated area so that they can run and exercise during the long winter months, another reason for leaving elephants and other large animals in the zoo where you can enjoy them without having to care for them.

Animals that spend a great deal of time in the water, such as capybaras, tapirs, seals, and otters, will need to have heated pools for the colder months as well as heated quarters. If you are determined to keep such animals, build a tropical, greenhouse-type structure for them, complete with pond and waterfalls. Make certain that it has a good drain, and also consider filtration of the water, should this be needed. Swimming pool filters as well as heaters will prove useful here. With today's modern styrofoam or fiber-glass insulation it is feasible to construct quite a large building with a dirt floor that can be planted with all kinds

The trumpeters are very tame and curious birds. Here they investigate a cigarette lighter, tossed on the ground by Bob Austin.

of tropical trees and shrubs. You could make a little Garden of Eden right in your back yard and stock it with small exotic birds and animals carefully selected for their compatibility. If you did most of the work yourself, you could construct a cement-block building fifty- or one-hundred-feet long with eight-foot-high walls and a high-pitched roof for just a few thousand dollars. The roof should be constructed largely of clear fiber-glass panels, some of them on hinges (screened of course) so that they may be opened in warm weather to allow the heat to escape. The cement-block wall should be insulated against the cold. A small heater will keep the inside temperature warm and comfortable all through the cold winter, and you will have your own private tropical wonderland. If you keep such pets as a monkey or an ocelot, they are certain to be much happier in such an enclosure than they would in any cage. You would of course need to select trees and shrubbery strong enough to hold the weight of the animals that you plan to acquire, but that is no problem. You could even provide a hollowed-out log in one of the trees for those animals or birds that would desire it.

If you have limited funds, you would of course make a much smaller building, perhaps something ten-by-ten feet, or purchase one already built and pay for it with monthly payments. On the other hand, if you are affluent and money is no object, you could make your Garden of Eden into an acre or more. One famous bird breeder covered ten or fifteen acres in the northern part of the country with glass and plastic and kept his birds in green, tropical habitats in an oasis of warmth while temperatures dipped far below freezing outside. If the water table in your area is very low, you might consider digging a trench ten- or fifteen-feet deep with a bulldozer or backhoe. Then all you have to do is run a roof across the top, either a high pitched roof or a flat one. In either case, it should be made of clear plastic and vented so that sections may be opened to regulate the temperature as needed. Clear plastic is better than glass because of the breakage factor. If you are keeping exotic birds or animals in your enclosure, a broken pane of glass could mean the escape of all your stock. Clear plastic window panes or translucent fiber-glass panels would be better and cheaper in the long run. You would probably need to reinforce the walls of the ditch with wire screen or even concrete to keep them from caving in and to keep the animals from digging out. You would also need to consider surface drainage from rain and melting snow. Perhaps a fair-sized pond at one end would be in order with an automatic sump pump to pump water out of the pond if it exceeds a certain level. You would have to put down a layer of rich soil on the bottom of the trench so that you could grow trees and shrubs in profusion. A few large rocks would add to its beauty. The area would also have to be heated in the cold months for maximum enjoyment.

If you do not want to bother with heated quarters or greenhouse-type structures, you can still keep some exotic animals by simply choosing those which can live in a cold climate. Animals like the bison, moose, wolf, Siberian tiger, clouded leopard, snow leopard, musk-ox, and even several species of monkeys can all live in the cold weather without heated quarters. But even these must be provided with a windproof and dry sleeping area so that they can get out of the weather from time to time. Their sleeping quarters should be dark and secure, preferably up off of the ground in the case of cats. It was found by zoos that many animals would not breed in captivity simply because they were deprived of privacy. Few animals, except for hoofed stock, have their young out in the open. They need a den, cave, or dark, sheltered area if they are to reproduce and care for their young. Although many tropical animals can adapt to a certain amount of cold, even freezing weather, this does not mean that they can be kept out in the snow, except during the warm part of the day and only for certain species. Polar bears, musk-oxen, moose, and other animals that live in cold climates can be left out in the coldest weather with certain reservations.

Naturally, it is easier to keep exotic animals in the warmer climates such as Florida, Southern Texas, and California. The man in the South can create his own African veldt or Australian outback simply by encircling a plot of hammock land with an eight-foot fence, taking care that an additional two or three feet of fencing should be extended underground to keep dogs or other animals from digging under the wire. The introduction of acacia or baobab trees or a dense strand of bamboo if the area is well watered can create a real jungle atmosphere and a happy home for carefully selected animals or birds. If you have a vast tract of land, say one hundred to several thousand acres, you may wish to set up a private big-game preserve and even include a herd of elephants. This would involve a considerable

A marmoset in the author's yard. He kept several of them in a large, tree-planted cage.

sum of money, since elephants are the most difficult of all animals to confine because of their enormous bulk and strength. One of the few fences that will hold them is made of railroad rails. The rails are cut into fourteen-foot lengths and sunk into the ground to a depth of six or seven feet. It may be necessary to anchor them in cement if the ground is soft. Holes are drilled through the rails above ground, about a foot apart, and one inch steel cable is strung through the holes and welded in place. This is an elephantproof fence and will restrain even a full-grown animal. Smaller elephants can be restrained with a much cheaper fence, electrified or surrounded by a deep ditch or moat if needed, but if you plan to set up a permanent elephant preserve, the stronger fence will be needed when they mature. Needless to say, anyone who has his own private herd of elephants roaming about on his land will have an exclusive, at least in the neighborhood. I would recommend a very large tract of land for such a project, and it should be divided into two sections so that when the animals have denuded an area of trees and brush, a gate can be opened to lure them into the second area so that the original area may be cleaned and replanted.

Who would want a herd of African elephants roaming about on their land? I would for one. I think it would be fabulous. You could construct a tree house in one section where guests could stay to watch the animals cavort about a waterhole or salt lick as they do in Africa. Or you could set it up as an attraction and charge admission. Many people would love to see elephants out in the wild, completely free and uncaged. Elephants are the chief attraction at the famous Treetops Hotel in Africa, which I visited while on safari. You could throw in a few giraffes, rhino, zebras, and other animals and soon you would have an outdoor zoo of your own. But it is only for the wealthy, for such a project would cost hundreds of thousands of dollars unless started on a very small scale. You will find much more about such big-game projects as you read further, for it was just such a project that led to the accumulation of information that enabled me to complete this book. Some of the world's most experienced animal handlers and breeders advised me with this work.

The proper wire and mesh size must be considered when building a pen or holding area for your prospective pets. Ordinary two-by-four wire fencing is all right for some large animals, but for many it is not satisfactory. Birds and large reptiles or small animals often cut their jaw or face on the wire when they run against it in an effort to escape. The wire opening is just large enough to allow them to stick their jaws through it so that they keep trying to force the rest of their body through the opening, sometimes cutting their face severely in the process. The small one-inch-by-two-inch welded wire is better because it eliminates this problem. Or even the smaller mesh of one-half-inch-by-one-inch may be preferable in the case of marmosets or small birds and mammals. The welded wire is much more expensive than the common chicken-coop wire but it is much more durable and will last five to ten times as long as the cheaper wire.

Chain link, of course, is suitable for many animals. It is widely used in zoos, and the newer open-air zoos are using it by the mile. It comes in different gauges, so one must consider not only the mesh opening of the wire but the thickness of the wire itself. Large animals, of course, will require the heavier gauge wire. The fence posts also will have to be strengthened accordingly. I always chuckle when mentioning chain link for animal enclosures. A few years ago, while trying to raise funds for my own big-game preserve, I wrote to a number of zoos for advice. I outlined my plans, mentioning that I would be confining many of the animals with chain-link fencing. Most of the zoos were enthused with my project, but one large zoo wrote back that chain-link fencing was not suitable at all and that they did not approve of my plan. The chuckle came when I found out recently that *they* are building a big-game preserve very similar to the project I had presented and that they are using chain-link fencing throughout! (I still have their letter—perhaps someday I shall send them a copy of it.)

One thing to remember with chain link if you are confining large animals like zebras, or buffalo is that the wire is placed on the inward side of the posts, facing the animals. In addition, the fencing should be securely anchored to the posts with strong, twisted wire (not soft aluminum). Should your animals become frightened and run into the fence, if the wire is on the outside of the posts, the animals will simply push it to the ground and escape. Also, remember to round off corners in the fenced area so that the animals will not stop in the corner if they follow the fence. Round or oval paddocks are better than square or rectangular paddocks because they eliminate all corners completely.

There are other things to consider when contemplating the acquisition of a wild creature, and this could be the most important reason of all for *not* keeping the animal: Is it fair to the animal to keep it in your house or yard, away from its own environment and forever without companionship of its own kind? Here one must search his own conscience for the answer. Can you imagine the heartbreak and loneliness that a monkey must endure when torn from its mother and confined to a tiny cage without ever again seeing members of its own species? Just put yourself in its place for a little while. Just imagine that you were captured by a pack of baboons and thrust into a cage for the rest of your life. You would have no one to talk to, and even if you were brought life-sustaining foods by your captors, you would live in fear and have a miserable existence. But if they put a gorgeous blond in the cage with you things would be looking up and your confinement would take on a brighter look, for the loneliness and boredom, at least, would largely vanish. Then if they built you a beautiful home with a clean bed, fresh running water, and a large outdoor area to explore and play, your confinement away from the rest of your race would at least be tolerable. The same, of course, applies to animals, particularly monkeys and other animals that live in a family group. If you keep them by themselves, you are condemning them to an unnatural life of utter loneliness. They may appear to be happy and contented with you when they are small, but when they mature it is only natural that they will want to have a mate and raise a family.

On a recent visit to the famous Monkey Jungle near Miami, the owner, Frank V. DuMond, explained to me the very close family relationship of a baby monkey with its mother and said that he did not think that monkeys should *ever* be kept as pets, for when the baby monkey is taken from its mother, the little animal loses a link with its species that it can never obtain by living with humans. I can see his point, but then what about the people who really *want* a pet monkey? Should they not be allowed to have one?

Perhaps the answer would be to sell monkeys or other animals only as mated pairs, or at least as male and female if they are getting along well together when sold. This may be a more humane way to keep a gregarious animal instead of keeping it all alone for the rest of its life. It is bad enough to confine a wild animal, but to deny it company of its own species is even worse. I feel that people who keep wild animals for pets or otherwise should always obtain a mate for their pet, from the very beginning. Then they will not only be providing company for the animal but there is a good possibility that their pets will

Bob Austin's beautiful, fenced backyard allowed plenty of room for his pets. Note Tarzan the chimp by the base of the tree.

produce offspring, which would help to insure survival of the species. In fact, if totally restrictive laws are passed against the keeping of all exotic pets, I do hope that an exception will be made for those who would make a real effort to raise the exotic animals. This could be an answer to the whole problem, for if the animals are raised in captivity, particularly through several generations, the pet owner will be getting an acclimated, superior animal, and yet he will not be depleting the stock in the wild. This could be true of nearly all species of wild pets, even those on the so-called endangered-species list. If enough people raised the animals in captivity, the price of the animals would go down and trapping of the species in the wild would greatly diminish.

Take the ring-necked pheasant as an example. This is an exotic bird that was brought to our country from Mongolia for hunting purposes. The first birds sold for five hundred dollars or more until people started to raise them in captivity. Now they can be purchased in thousand lots for as little as *fifteen cents* each. The same thing could happen to exotic pets of all species. If enough people raised ocelots, cheetahs, and monkeys of all types, the price of all of them would go down so much that it would no longer be profitable to hunt them in the wild. It would appear to me that this would be a far better conservation effort than simply banning the importation of the animals. I am firmly convinced that practically all exotic animals and birds could be raised in captivity if proper housing were provided for them. I feel that if a natural area of trees and brush were simply covered or fenced and proper hiding places or houses were provided, then most animals would thrive and reproduce in captivity. This would also apply to reptiles and birds, even to fish. The idea is to make the environment as natural as possible so that you can closely duplicate the area in which the animal naturally lives. The cage or area of confinement should be as large as possible, depending, of course, upon the size of the species to be kept. You should never keep an ostrich, rhea, cheetah, or antelope in a small cage or backyard. They need large areas in which to run at top speed if they are to live and develop properly. Nor should a monkey be kept without trees and branches upon which to climb. Water animals should have clean water at all times, and if you live up north, tropical animals must have warm quarters, year round.

Learn all about your prospective pet before you get it. Buy a book on it and read about it so you will know about its natural habitat, and try to duplicate this habitat as closely as possible. Do not buy one animal—buy *two,* a male and a female, if you do decide to adopt a wild friend. But do not buy any at all if you must keep it in a small, wire cage. Think of how you would feel if you were cooped up in that cage day and night for the rest of your life. Then perhaps you will get a cat or a dog instead. Also, remember that all baby animals will grow up and that they will not be as cute when they are big. There are numerous instances of lions attacking and even killing their owners. There are also cases where they have escaped and killed children or adults. Would you want this on your conscience for the rest of your life? Remember, all baby animals are adorably cute but do not buy one on impulse. It may bring you far more heartache than happiness.

In the following pages, I will relate my actual experiences in keeping a number of exotic creatures. You will see that although it was interesting (to say the least) it also presented problems, and tragedy was not lacking, which is often the case. The main lesson I learned was that wild animals should have a large and natural enclosure if they are to survive and be happy. They should not be kept any other way.

3
An Ostrich in the Yard

At one time, our backyard bore a close resemblance to a zoo. Not only did it contain the usual dogs, cats, chickens, and rabbits, it also housed monkeys, capybaras, wallabies, turtles, giant lizards, and three ostriches. Fortunately for us, the ostriches were not the huge African species that would have soon outgrown our yard. They were the South American variety known as the rhea, a smaller member of the ostrich family, of similar shape and habits.

We obtained the rheas as chicks from Alton Freeman and Bill Chase when the birds were about the size of domestic chickens. They grew rapidly, devouring huge quantities of food right from the start, and we found it increasingly difficult to keep them filled up. One of the birds cried a great deal. It appeared to be the weakest of the three birds and was much more shy than the others, but later we found that this was the male, for only the males have vocal chords. When the birds were small, we let them run loose in the yard. They quickly became very tame, accepting food from our hands. They ate a variety of foods, mainly chopped fruit and Purina Dog Chow. We mixed their food with small quantities of coarse sand, which was suggested by Tony Gomez, Bill Chase's associate, as necessary to provide grit in their diet. But they also required huge quantities of fresh lettuce, which presented a slight problem. At first we fed them from a small dish that served the purpose, but soon we had to use a large Easter-egg basket and eventually a washtub. It became a gargantuan task just to supply them with lettuce, and my wife had to bring home a huge box of lettuce leaves from the grocery store each week in an attempt to ease their insatiable appetite. The more they grew, the more they ate. We soon had to buy their food in fifty-pound bags. In addition to all the food we gave them, they soon ate all the flowers in the yard.

We finally had to pen them up at night, but the first thing they did when we turned them loose in the morning was to head for the flower gardens, where they would devour every flower in sight. They not only ate the flowers, they ate the entire plants.

To watch a rhea run is pure delight, for they are no doubt among the champion runners of the world. Our three ostriches would run around and around and around our house, picking up speed with each passing, until they were running so fast they would almost become a blur. They would chase after each other with wings outstretched for balance and would run in an erratic fashion, darting quickly from left to right and changing directions abruptly as they ran. Like their African relatives, they are true ground birds and cannot fly. Running is their only means of defense, except for a strong, downward kick from their powerful legs, which they use only as a last resort.

The young birds did not have the balance and coordination that they would attain as mature birds, and would often run headlong into each other at top speed or lose balance completely and go hurtling across the yard in a wild melee of long legs and twisted necks. Sometimes they would run around the house in opposite directions and end up in one huge pile as they all collided. It was hilarious to see them run. Nearly anything would start them running and they always ran at full speed. If one of the children suddenly ran into the house, the ostriches would see him running and decide they should run too. Apparently this was their natural instinct, for they probably felt that if anything was running, something must be chasing it and they did not want to be around when it got there. Even passing cars or trucks would make them start on their fantastic, erratic dash. They seldom ran straight ahead. It was always a wild, evasive dash, weaving from side to side and, of course, at top speed. It was very comical, to say the least.

A pair of rheas that the author raised from tiny chicks. The female is on the left. Note the thickness of the male's neck compared to that of the female.

The ostriches grew fast, and soon they were over four feet tall. They began to draw traffic to the area, and it was a common sight to see cars parked in front of our house while the occupants craned their necks to get a glimpse of the huge birds. One day I noticed a telephone truck parked in front, and the astonished driver was staring into our yard as though he was in a trance. He stayed there about a half hour and could not believe his eyes. I guess he was not quite accustomed to seeing ostriches in someone's yard.

One quiet night we heard a tremendous, loud, booming voice emanate from the backyard. It was so loud that it was frightening. We could not figure out what it was. A few nights later, we heard it again. Surely the ostriches could not make it. Or could they? They could! One afternoon I heard the noise again and it came directly from the ostrich pen. What a blast! After that we heard the noise with increasing regularity. It was the mating call of the rhea. We had noticed that, as the birds matured, one of them had a much thicker neck than the others, and after reading more about the birds in numerous books, we came to the conclusion that this was the male. Now he was serenading us!

Our conclusion proved correct, for this bird grew much larger than the others and eventually became very mean and aggressive. We found very little information about rheas in any books and had to learn about them ourselves. The female birds remained tame and would take food from our hands, even when they became adults, but the male became more vicious every day. Instead of taking food from our hands it would intentionally grab the hand instead. It grew much larger than the females and was clearly their boss as well as protector. I remembered that when it was a chick it was the most frail of the lot and we wondered if it would survive. Now it was robust and powerful.

We had to get rid of one of the ostriches because it developed leg trouble, which we suspect was caused from running into a fence at top speed. This particular bird had never adjusted to life in the yard and was extremely restless. One day it shocked us by jumping over a six-foot fence. Everyone had told us that the birds could not jump at all, and most of the fence in our yard was only four feet high. Yet this bird had cleared a six-foot portion of it with apparent ease. The bird simply hopped over it. It worried us a great deal for I

First stop in the morning would be Mrs. Straughan's periwinkles near the back porch. As the ostriches grew, they ate nearly all of the flowers in the yard.

could just picture myself chasing an ostrich down the road with all the neighbors looking on. I probably would not be able to catch it anyhow and it would end up in downtown Miami. But fortunately for us it did not jump over the lower fence. My friend Warren Prince finally took the bird away with him for examination by a veterinarian at Silver Springs. This left us with two birds that luckily turned out to be male and female.

I had given the birds a large area in the back of our property so that they would have sufficient room to run around. It was as good or better quarters than they would have received in many zoos. It was open grassland with a few small trees and shrubs. My whole purpose in acquiring the rheas or ostriches in the first place was to attempt to raise them, so I made them as natural a habitat as possible. The pair reached their full size in about two years from the time I had obtained them. They settled down to a regular routine, and my ostrich project appeared to be going well.

One day as I was bringing their food to them I was flabbergasted to find that the male bird had constructed a huge nest. He had hollowed out an area in the ground and had gathered together huge quantities of grass and leaves, which he had worked into the nest. He sat on the nest for hours at a time. The female paid little or no attention to it.

An ostrich (rhea) in the author's front yard. The birds' joy was to run around the house at top speed. They would run so fast that they would almost become a blur. Needless to say, they attracted a lot of attention.

Next, the male began an elaborate courting of the female. He would approach her with wings outspread and move his head up and down in a circular motion. He did this all day long, for days on end and we wondered if his head would fall off. The female paid little or no attention to him, but we did notice that her body was becoming very fat, and we suspected that she was carrying eggs. One day I noticed that she was sitting on the nest. This was good news. The male had finally gotten the word to her that the nest was for *her* after weeks of head swinging and wing fluttering. We thought it would have been so much simpler if birds could talk. He could have told her the first day that the nest was for her, instead of going through his endless strutting, wing fluttering, and head swinging. It would have saved him all that work.

The rheas at rest on the lawn. Sometimes they would stretch their neck out in front of them or sprawl with their feet in the air as if they were dead.

The male gave out with his loud, booming mating call with ever-increasing frequency. It sounded like a combination of a foghorn and a horse. It was very loud, especially in the quiet of the night. One neighbor said that the first time she heard it she nearly jumped out of her skin. We were worried that some of the neighbors might complain, because the sound could be heard for at least a half mile. The male became extremely aggressive now, and if anyone went near the fence, it would charge at him at full speed, hurtling its body against the wire and nearly knocking the fence down. It had grown into a very

The full-grown male rhea. The male is distinguished from the female by the thicker neck, which also contains its vocal cords. Only the male can make a call, which sounds like a sighing horse. This big fellow was more than five feet tall and grew meaner by the day, chasing the author about the yard.

powerful creature. I estimated its weight at between seventy and ninety pounds.

One day I had to go inside the enclosure for something. I had sneaked inside while the bird was not looking, but it caught a glimpse of me out of the corner of its eye and attacked. I had no time to get out, and the big fellow backed me right into the corner. It was only five feet tall, but when it ruffled up its feathers it stood very high and looked down at me with menacing beak and thrashing feet. I felt very insecure and a little foolish to be made so helpless by a bird.

Fortunately I had a large, plastic wastebasket with me, so I crammed the basket down on top of its head and pushed it to the ground. I found this to be the best protection for an ostrich attack. If you cover its head, it cannot see what it is doing, and it is rather helpless. When I finally let the brute up off the ground, it attacked me again, so I crammed the bucket over its head once more. It finally got up and ran away from me. Thus I was spared the cruel fate of being stomped into the ground by my own ostrich.

I had cautioned the children about going into the ostrich pen, but one day my daughter Julie disobeyed and went inside to collect feathers. She had just reached the far end of the pen when the big male saw her and took off after her at top speed. She was screaming with fright and ran for the exit with the huge ostrich chasing after her like a chicken after a grasshopper. It caught up with her just as she reached the fence and gave her a good peck on top of her head. She climbed over the fence in record time. I told her that she got what she deserved for disobeying, but I am sure the peck from the big bird hurt little more than her pride. She was very small at the time, and I feel that it did not go after her with the same vengeance that it would have shown toward an adult.

A rather humorous incident regarding the ostriches occurred with one of Julie's teachers at school. Julie had been reprimanded for something by her first-grade teacher, and she told the teacher that she would like to put her in our ostrich cage and let the ostrich eat her. The teacher told us about it (with a good natured laugh), then asked, "You don't *really* have an ostrich, do you?" We told her that we had two of them!

After more weeks of head swinging, wing fluttering, and loud booming noises by the male ostrich, the inevitable happened. I went out to see my birds one morning, and there on the ground was a huge white egg. It was just lying out in the open and was nowhere near the nest. Neither bird was paying any attention to it, but as I approached the pen, the male went after me with a wrath and fury that knew no bounds. It would rush at the fence heaving its body against it, making a loud hissing noise while trying to reach over the top of the fence to bite me. It was absolutely enraged. I decided to leave well enough alone and retreated back into the house to spread the news. We watched the birds at various times of the day from the window. The female paid little or no attention to the egg, but the male was definitely interested in it. It would roll the egg around with its bill and then would go back to sit on its nest. We thought that the egg should be *in* the nest, but further reading and inquiring among bird experts told us that the egg should not be in the nest for the time being. The birds would lay their eggs in the general area, but they would not sit on them until a certain number had been layed, otherwise the birds would all be born on different days. This sounded logical, for I had remembered when raising chickens that the hen would not sit on just one egg. It would wait until there were a dozen or so eggs before sitting on them.

The lone ostrich egg stayed out in the hot sun for several days, and then the female layed another egg. This went on for weeks until there were fifteen to eighteen eggs altogether. Meanwhile, the male bird had made another nest at the opposite end of the yard. By now, the eggs were scattered all over the yard. The male had pushed a couple of the eggs into a nest, but then it went over and sat on the empty nest. This was frustrating and I began to doubt its wisdom. Finally I bought an incubator and decided to try to hatch out the eggs myself. I collected about twenty eggs (each one a life and death struggle) and left a few out in the yard in the event that the birds would sit on them. But neither bird sat on the eggs and instead seemed to totally neglect them. Meanwhile, I carefully turned the eggs in the incubator each day, and after a few weeks I candled the eggs. None of them were fertile, so of course none of them hatched. This is probably why the birds did not bother with the eggs. The big male was not such a dumb dodo after all.

I talked with Alton Freeman and Bill Chase about the eggs and they both agreed that perhaps the birds were too young to have fertile eggs and that I would have to wait a year or two for both birds to mature. I was disappointed in not being able to hatch out any of the eggs, and the male ostrich was becoming more of a problem. It became so aggressive that it was actually dangerous, and I was fearful that it might hurt someone. It grabbed my hand one day while I was feeding it and inflicted a painful laceration, which did not endear it to me very much. The downward kick of the large ostrich is its greatest potential danger, particularly to humans. It can slash downward with lightning speed, and its sharp, pointed toes can inflict serious injury, ripping open a piece of flesh like a saber. A large bird like the African ostrich could easily kill a person in this manner, and it is possible that the rhea could do the same under the right conditions. It is said that the only defense against an ostrich out in the open is to lie flat on the ground. Then they cannot kick you, but you must protect your head from their powerful beak. Though the South American rhea is not nearly as large as its African relative, an enraged male is nothing to trifle with.

I was very disappointed in my failure at hatching the eggs, but rather than throw the eggs away, I drilled holes in them at the top and bottom and flushed out the insides by holding the end of a garden hose to the opening at one end. It worked perfectly and was much better than blowing through one end to clean them. Each of my children now have a cleaned ostrich egg for their "museum," and we gave the remainder to friends. Meanwhile, the male ostrich had become so aggressive that I decided to sell both birds before someone was hurt. But unfortunately, tragedy struck before I could act. I came home from fishing one day to find the female dead. Apparently the male had killed it while mating. It was most likely unintentional, since the two birds seemed to get along well. It left us with just one bird, the totally mean male.

I felt very bad about losing the female, since I had raised it from a chick, so I decided to have it mounted. The only box I had that was large

The male courts the female by spreading his wings and following her about the yard. He also calls to her with a loud, booming voice.

The male rhea would court the female by spreading his wings and following her about the yard. He built a huge nest in one corner of the yard, and would fiercely defend it. Dozens of eggs were laid, but unfortunately none hatched.

enough in which to pack the body was a large, live-tropical-fish shipping carton. I pushed the huge body into the box, but rigor mortis had set in and the huge legs were stiff. They would not bend at all, so I closed the box with the huge legs sticking out from the top. It looked rather strange, so I put a large paper bag over the legs. Then I rushed down to the bus station to ship the box to the taxidermist. When the clerk asked what was in the box, I simply stated, "An ostrich." He looked at me as though he doubted me, but sent the shipment on to its destination. A few hours later, I called the taxidermist to ask if he had received the ostrich without any trouble. He broke out into a laugh. He said that the paper bag had fallen off or had been removed, and it was a strange sight to see the huge legs sticking straight up from the box in the middle of the crowded bus terminal.

We finally sold the remaining bird, and although I was glad to see it go since it was so mean, I felt sad about my ostrich venture. Evaluating the situation now, I would like to try to raise the birds again sometime, perhaps even their African relatives, only this time I would do it on a larger tract of land of twenty-five or fifty acres where they would have a secluded place in which to run and nest. I feel that birds of this size should have a large area. A small backyard is just not suitable for them. They should also have a small pond, because they like water and will even swim if it is deep enough. Also, their fence should contain

The nest of the rhea in the author's backyard. It was constructed of grass and weeds and was about three feet wide. After building the nest, the male would call to the female day and night in a loud, booming voice. It would also strut about the yard with outstretched wings and sway its head back and forth in a wide swing for hours on end. One day, the female laid one huge egg, then another, until more than two dozen had been laid.

no corners. The corners should be rounded like an oval so that when they run along the fence, they will never come to the end of it. Also, they should be confined with an eight-foot-high fence with narrow mesh chain link or wire. Welded wire of one-by-two-inch mesh would be best. I found that with larger mesh wire, the birds would cut their bills very badly as they pushed their heads into the wire, either when attacking or when frightened.

Rheas make fascinating pets—when they are young—and they are most amusing to watch as they run wildly about the yard. But remember that they will grow up fast, and then they will present many problems, as I learned from my own experience. I think that a full acre of land would be suitable for a pair if it had a few small hammocks of trees and brush so they could have seclusion for nesting. It would have to be out in the country in the back side of the land so that cars, passing trucks, or the neighbor's dogs will not frighten them.

An ideal enclosure for a few rheas, or even African ostriches, would be a ten- to thirty-acre pasture with some hammock and brush for privacy. Properly fenced with rounded corners, it would make a fine home for them, and a few compatible animals could be kept in the same compound. You may have to enter the area in an enclosed jeep or land rover, especially if you are keeping the larger African species, because these can be very dangerous. Keep in mind that if you want an ostrich for a pet, either have a very large backyard or, from the very beginning, decide beforehand that you will only keep it for six months. After that it will present you with a great many problems, the least of which will be an enormous feed bill. Both ostriches and rheas are in demand by zoos and tourist attractions, so that should you succeed in raising them, it would probably be a profitable venture, especially if you live in a warm climate and put them out to pasture on a huge tract of land. Not only will this save on the food bill but should the demand for them be slow, you could keep them out to pasture until there is a market for them.

At this writing, even the unhatched eggs of the African ostrich have been offered for sale at three hundred dollars each. I have seen the cleaned and polished eggs for sale in our local department store for as much as sixty dollars. In addition, the feathers of both African and South American birds are worth a considerable sum when there is a market for them. A full-grown African ostrich is worth close to a thousand dollars at present. But do not try to raise either species without the proper land. It will only be a disappointment. There is nothing funnier than an ostrich running around your yard, except perhaps when it is running after its owner who is frantically trying to get away!

4
Jumping Wallabies!

There is nothing quite as cute as a small kangaroo hopping around your backyard. Or at least that is what I thought when I ordered three of them from the London Zoo. I was working on plans for a big-game preserve project here in Florida and decided to get a collection of animals together for the project. A large landowner had promised that if I could raise the funds for the animal preserve, he would let me use ten thousand acres or more of beautiful veldt-like land for my big-game project.

To build such an animal preserve had been a childhood dream, for I had always believed that America should have a genuine, authentic jungle, complete with herds of elephants, giraffes, hippos, and rhino so that the millions of people who could never go to Africa in their lifetime could see essentially the same thing right here in our own country. It was, admittedly, a grandiose plan, and I do not pretend to be the first to have thought of it. There have been jungleland projects in America from time to time in the past and there are many of them operating today. The major difference between my project and others was that I planned mine for a vast expanse of land with a minimum of ten thousand acres on up to several hundred square miles. I reasoned that it was more practical to raise kangaroos or giraffes on a piece of land than to raise cattle. Both eat grass, but the difference is that while a cow may be worth only a few hundred dollars, a giraffe is worth nearly five thousand dollars. Furthermore, you can charge admission for people just to go into the land to look at the giraffes roaming free.

Such thoughts prompted my first kangaroo order, or rather wallaby order, for this is more correctly what the animals were. I had made contact with the London Zoo and arranged for the wallabies. It was an expensive deal, and when the wallabies finally arrived, I estimate that I spent close to a thousand dollars for three of them.

The species I obtained were Bennett's wallaby, a medium-sized animal about three feet tall. They are one of the most beautiful members of the kangaroo family, with rich, dense fur and gentle, intelligent faces. It was an exciting day when they finally arrived at the airport after a lengthy correspondence, telegrams, phone calls, and months of preparation. My good friend Warren Prince had accompanied me to the airport and helped me clear them through customs, which was no easy task since they had come from a foreign country. I had known Warren since early childhood, and he shared my keen interest in animals. In our early years we had both collected small animals in the wild in New England and both had kept all types of pets and wild creatures most of our lives. Warren helped me load the crates with their precious cargo into my station wagon, and in a short while we were home, preparing to release the animals in a quarter-acre enclosure that I had selected for them. We placed the crates up to the fence and opened the lids. Soon, out hopped two fine, healthy-looking wallabies. We stepped back so as not to frighten them, and soon my backyard had an Australian appearance with real live "kangaroos" hopping about the place. At the time, wallabies were quite uncommon in our country. Many zoos had none at all. It had taken a great deal of effort to obtain my specimens.

The wallabies settled down in the yard and appeared to be doing well. A few weeks later, I

Little Sunshine, a baby wallaby born in Miami in the author's private "zoo."

acquired a third specimen from the same source. They all got along fine together, and many people came to see them, particularly children, for even our local zoo did not have any at the time. They were delightful animals. They do not make any noise and are grazers, making the keeping of them a relatively simple matter. They would hop about the yard like giant rabbits and stop to wash their hands every half hour or so. But they were not tame animals and were easily frightened. If they were approached too closely by anyone, they would run into the fence in the opposite direction at top speed. This caused considerable worry, for they could cut themselves on the wire fence in their anxiety to get away. I had to caution everyone about approaching them too closely. Our own dog would terrify them when they saw it, even at a great distance. They would flee in utter panic.

Weeks went by and the wallabies settled down somewhat. I visited their compound many times every day to observe them. Then one day I noticed a slight wriggling in the pouch of one of the female wallabies. At first I thought it was my imagination because it was ever so slight, but when I saw it several days in a row, I knew that one of the wallabies was pregnant. Then I noticed the same thing in the other female and my excitement knew no bounds. I began to envision dozens, even hundreds, of wallabies hopping about the yard and eventually hopping about the Australian area of my big-game preserve. Weeks went by and the swelling in the pouches grew, until one day as I was watching them, I saw a tiny head poke out of the pouch and look straight at me. I was flabbergasted! It was a baby wallaby. I called my children and they rushed outside to see it but they were too late. The little guy pulled its head back inside so no one could see it.

Next followed a daily vigil of hiding behind trees and peeking out into the yard to get a glimpse of the elusive baby wallaby. But finally we were all rewarded with a quick look at the sleepy little fellow who would stick its head out of the pouch for a quick look at the outside world and then duck back quickly back into the warm pouch. Friends of the family would make a special trip to our house just to get a quick look at the new baby. If they were lucky, the little head would poke out so they could see it.

But all did not go well with my wallaby clan. Apparently they had developed a virus or pneumonia, for one day I noticed that one of the females was coughing badly. And alas, just a few days later it was dead. It was a sad event but one that inevitably happens to those who try to keep exotic species from other lands (including zoos). The other female looked in good health but I was worried that it would come down with the same disease. I had shown the deceased animal to a local veterinarian who came out to our house to inspect it. He said that possibly the chill from the overseas flight had brought on the disease, but he offered no suggestions for treatment of the remaining animals. The events that followed the death of the first wallaby were equally sad, perhaps more so.

I observed the remaining female wallaby for several weeks, watching the little wallaby mature in its pouch. At first, all I could see was the tiny head peeking out from time to time, for when little Sunshine, as we had named the baby, was growing up, it spent most of the time snuggled up deep inside its mother's warm pouch. It was actually five months from the time that I had first seen a movement in the female's pouch to the exciting moment when the little fellow first stuck its head out of the pouch.

Weeks went by before the baby looked out at the world again. This time I noticed that the baby was not quite so pink as it was the first few times we saw it. Then I noticed that it was getting a little hair on its head. It stuck its head out of the pouch more frequently now, much to our delight. A baby kangaroo is really quite a creature. Its little ears are bent over and its nearly hairless face makes it look almost exactly like a Mexican Chihuahua. It is a completely helpless creature as a baby and stays in the warmth and safety of its mother's pouch until it is quite mature and ready to shift for itself.

What a pity, I thought, that in their native Australia both kangaroos and wallabies are slaughtered by the hundreds of thousands by sheep farmers who shoot the animals and leave them to rot by the road or bring them into town to be ground up for pet food. It would appear that in the vast wilderness of that country, where ordinary ranches sometimes comprise several hundred thousand acres of land, the government would require the sheep farmers to share the land with the diminishing wildlife instead of wiping it out to the point of extinction, as has happened in our own land. But perhaps this will come in time, for the conservationists are now being heard both in

The children brought the little fellow into the house.

our land and abroad. I do hope that the people of Australia will start to protect their interesting wildlife before it is too late.

We watched our little wallaby grow, and each time it stuck its head out of the pouch it was a special treat. As the weeks wore on, the little fellow's ears straightened out like its mother's and more hair grew on its pretty little face. Then one day, I surprised the little wallaby while it was hopping around outside the mother's pouch. When I approached, it looked furtively around and then stuck its head into its mother's pouch and quickly hopped inside out of sight. What a frail little fellow it was! I saw the little wallaby outside the pouch a few more times the following day, and one time when I approached, the mother ran off and left the little baby all alone. I picked up the spry little fellow and it appeared unafraid. I saw that it had a tiny pouch and was a girl. Very carefully I placed it on the ground near the mother and once again it jumped into her pouch for security and no doubt lunch, since it was still nursing.

The next day, tragedy struck. I went out to see the wallabies and a pathetic sight met my eyes. Mama wallaby was dead and the little wallaby was trying desperately to get inside the pouch. I swallowed hard as I picked up the little baby and carried it into the house. It was a shock, for the mother had appeared so fat and healthy and had every kind of food it could desire. I showed the little joey to my wife Rosemary, and it was such a bright and cheerful little creature that we immediately called it Sunshine. It was as bright and cheerful as the morning sun.

The children were at school when I brought the little wallaby into the house, but when they came home, Paul and Julie were just overwhelmed at the thought of a pet kangaroo in the house. Little Sunshine was a joy to behold. It had so much spirit and charm. It would hop about the house following anyone who called it and would make a strange little rasping noise in recognition when anyone spoke to it. We became very fond of the little charmer. It was so clean and completely odorless. We made a sleeping box for it with a heating pad on the bottom and put in a soft cloth bag in which it would curl up to sleep as a sort of a substitute for its mother's pouch. It was a heart-rending sight to see the little wallaby hopping bravely about the house as it followed the children from room to room, and we grew accustomed to the little feet padding about the house as Sunshine looked around for someone to hold it. Unlike most pets who would rather not be held, Sunshine never wanted to be left alone. It wanted someone to hold it up close all the time so it could feel their body warmth as it did when its mother was alive. During the day while the children were at school, if I was lying on the couch, I would hear a little scampering of feet and then see a little head peering up at me over the edge of the couch. Then little Sunshine would give a tremendous spring and land on my lap where it would curl up and go to sleep. Everyone in the family had a special place in their hearts for the little wallaby.

One evening, while my daughter Julie was playing on the rug with little Sunshine, the wallaby looked down the neck of her pajamas and decided that perhaps here would be a perfect place for sleeping. It promptly stuck its head into the pajama top and crawled inside. This was just like its mother's pouch, and Sunshine quickly curled up and went to sleep. After a little while, Julie took it out but it quickly crawled back inside, deciding that this was the best place for it. This went on for several days and the children took turns carrying the wallaby around in their "pouch." The children discovered that if they jumped around, they sort of imitated the mother wallaby and little Sunshine enjoyed the ride. It would curl up and go to sleep as soon as it was inside their shirt or blouse. But just when we were beginning to think that little Sunshine would survive without its mother, I noticed that the little wallaby would not eat its food or drink its milk. By afternoon, its tiny legs had stiffened and, despite all we could do, little Sunshine had joined its mother and departed from this world. It was a sad day for all of us, but for the short while that it was with us, the little wallaby had brought much sunshine into our lives. I have had many pets in my life but none so adorable as this cute little wallaby who captivated our hearts from the moment it entered our home.

I decided that I did not want any more wallabies for quite a while, but the next time I get some I hope to obtain them from a place within our country so that they will already be adjusted to our own climate. I finally sold the remaining male to an animal dealer after I had kept it by itself for several months. It appeared to be lonely and I did not feel that I would want to import any more of the animals after my last experience. The next time I try wallabies or kangaroos, I will get a trio of two females and one male, only this

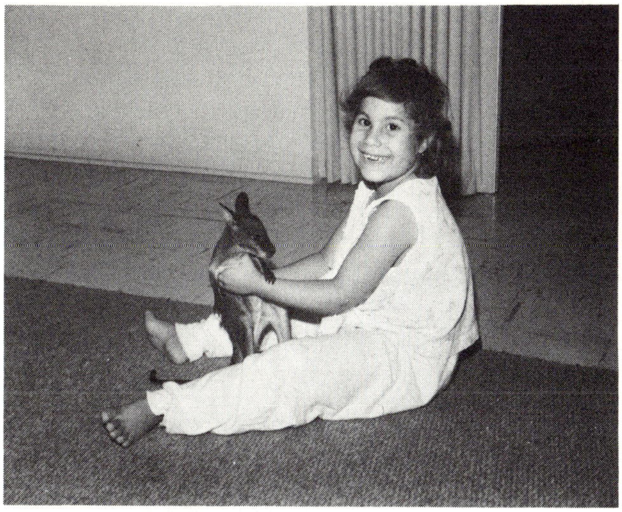

The author's daughter, Julie, playing with the adorable little wallaby in the living room.

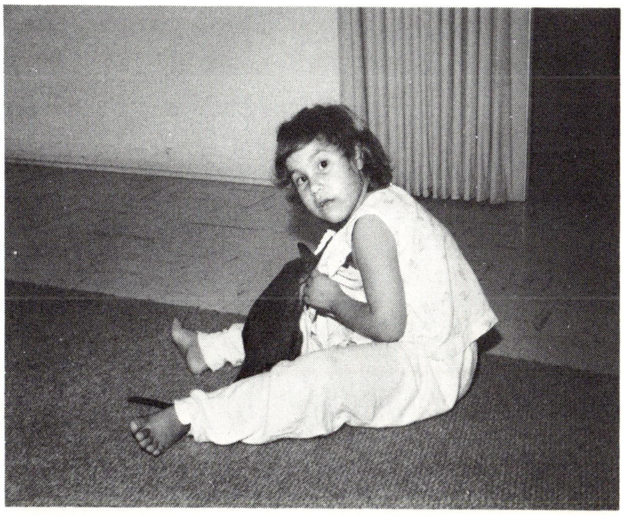

As Julie bent over, the wallaby discovered the opening in the top of her pajamas. It reminded the little animal of its mother's pouch.

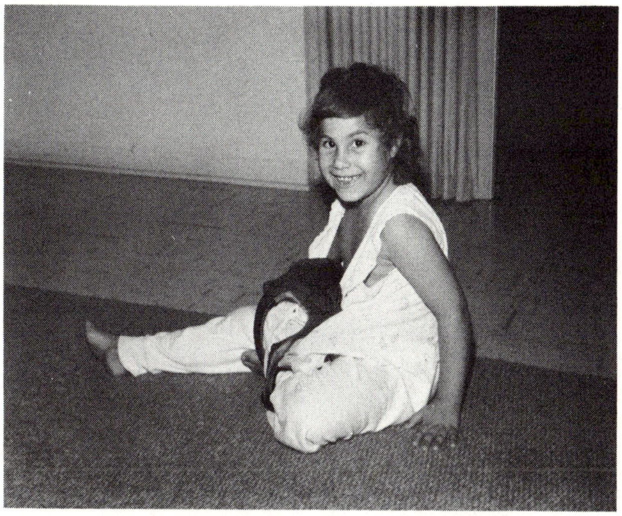

It crawled inside, much to Julie's delight.

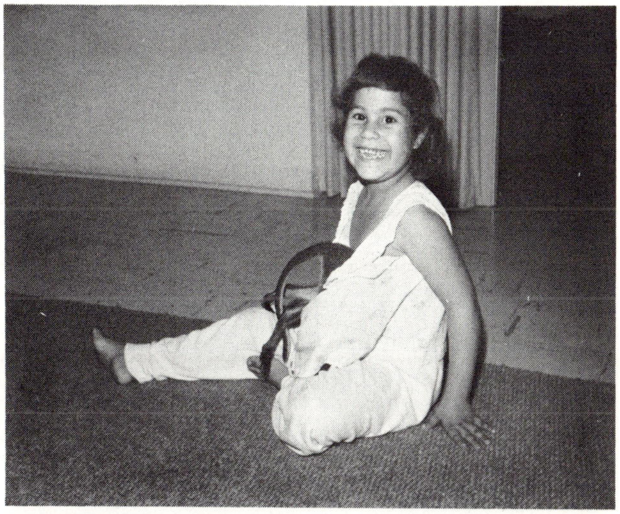

The wallaby crawled into the "pouch" and went to sleep.

Julie held the little wallaby tenderly. It was a sweet little fellow, soft and gentle. It wanted only to be loved.

time I will get some that have been in the country for a while so I can obtain local stock. Wallabies are excellent for keeping excess grass in check, and the London zoo reports that they keep them in large groups especially for this purpose. One would think that the highway department could use them for grass control in areas around bridges or overpasses where mowing the grass is difficult. It would give passing motorists a change of scenery.

Both wallabies and kangaroos are fast becoming rare and even extinct in their native Australia, where they are being killed by the thousands by ranchers much the same way that the bison were nearly exterminated from our own country. It would appear worthwhile to establish these animals in our own country. With man's ability to kill and exterminate such small creatures as the passenger pigeon and the Carolina Parrakeet, it would not be difficult to exterminate kangaroos if they ever did get out of hand in our own country. I disagree entirely with the purists who think that we should never release any foreign animals in our land because they might take over the countryside and multiply to uncontrolled numbers. In the very countries where exotic game has been released, most of it has ultimately helped the economy of the country rather than harmed it. Large animals such as deer, antelope, and Barbary sheep have been introduced to countries with complete success. Yet we never hear of deer overrunning a country except possibly New Zealand. We *do* hear of venison being on the table of many families who never had meat before. There is too much hogwash about the English sparrow and the mongoose. These same purists never mention the ring-necked pheasant or the rainbow trout, which have proven very successful.

I think that all species of wallaby and kangaroo could live out in the wild in certain parts of our country and would do less harm than one giant chemical plant spewing poisonous fumes into the air or turning the local rivers and streams into open sewers. Few animals are as cute and inoffensive as a baby wallaby. It is just too cute for words.

5
Miniature Dinosaurs

The strange, three-horned chameleon from Africa is probably the closest thing to a living dinosaur in the world today. It is a fascinating creature and in appearance looks somewhat like the triceratops of long ago. Yet despite its fearsome appearance, it is completely harmless. In fact, it moves so slowly it cannot even run away.

Here is an ideal pet for a child or adult who wants something interesting and unusual. Everything about this old-world chameleon is unusual. Not only is it strange in appearance but everything about it is bizarre and unorthodox. The three horns on its nose most certainly are not commonplace. Neither are the eyes, with their swivel sockets that telescope out from the body. The beady little eyes in the center of the turret will watch you with every move, and each eye moves independently from the other. The weird creature can watch your head with one eye and your feet with the other, both at the same time. Even the creature's feet are unusual. They look like tiny hands except that they open in the center, like a lobster's claw, when they wish to grasp a twig or branch. Just watching them walk is pure fascination.

The chameleon is never in a hurry. When he starts on a journey, the entire trip may encompass a total distance of only six feet. He may not even get that far. As you watch him walk with his slow, mechanical movement you begin to wonder if he is ever going to put his foot down to complete his step. He may hold his foot in mid air for minutes at a time as though frozen in the act of moving, while he contemplates his next step. This is a creature with no time schedule. Just watching him walk is worth the effort in keeping one.

But this is only a beginning, for the old-world chameleons are full of surprises. We all know that they can change colors. They do this whenever it pleases them, either to camouflage themselves in the foliage in which they are living or to regulate their body temperature. I have seen an entire group of chameleons all change their coloring to jet black in the early morning and then arch their bodies and expose them to the sunlight to absorb the maximum heat from the early morning sun. Then after they had warmed up, they would change to brilliant green to yellow to any combination of colors. They can so camouflage themselves that they can sit on a piece of bark for hours before you would even notice them. They are completely quiet and only when they move are you likely to distinguish them from their surroundings.

I have kept a number of chameleons of various species from Africa, Madagascar, and the Mediterranean area. I have had the three-horned species, both males and females, as well as giant species from Madagascar that were nearly two feet in length. They all became tame very quickly and could soon recognize a human approaching them with food. I kept most of my specimens in a large, outdoor cage that was well planted with trees and shrubs. They would eat all kinds of insects and were especially fond of crickets and grasshoppers, which they would eagerly take from your fingers. The larger chameleons had tongues that were easily six inches long. You could hold a grasshopper loosely, and they would snap it up from a distance of half a foot away. I should emphasize here that food should always be held lightly when hand-feeding them, for their tongue is not too well attached to their mouth. One time

The strange three-horned chameleon looks like a creature from another age. Despite its fearsome appearance, it is totally harmless. It walks in slow motion. Its feet have claws with toes on each side. It moves one foot at a time, sometimes holding a foot in the air for minutes on end until it finally decides to put it down and take another step.

while I was feeding one, I did not release the grasshopper fast enough and I accidentally pulled the poor creature's tongue completely loose from its jaws. The poor fellow went around without a tongue after that but he was still able to eat. I would stick the grasshopper directly into his open mouth and he would chew it down. It appeared that he was growing a new tongue but he crawled away from his cage one day and I was never able to verify it.

Watching a chameleon feed is another amusing and interesting reward in keeping these amazing creatures. They seem to be very intelligent and observant animals despite their lethargic movements. As soon as they sight a prospective meal, their eyes seem to brighten. It almost looks like they are smiling. Then they move their body into position and focus their eyes directly on target, almost as though they are aiming a gun at it. It takes them but a moment to draw a bead on their dinner. Then, quick as a flash, the long, sticky tongue darts out and engulfs the insect, pulling it quickly back into the mouth where it is slowly chewed and swallowed.

Chameleons will eat almost any type of hard-shell insect but are especially fond of crickets and grasshoppers. If you live in a warm climate you can keep them outdoors in a well-planted cage that should offer them both sunshine and deep shade, as well as protection from excessive rain. They cannot stand the hot sun very long, so some deep shade must be provided for them. They usually like to sun themselves in the early morning after they have awakened, and when their body temperature has reached the proper level they will move off into the shade to search for food or possible romance.

The little "dinosaur" looks at the author with its strange, telescopic eyes. This three-horned chameleon comes from Africa and looks somewhat like a small triceratops. An exhibit of old-world chameleons, each in a richly planted terrarium, would be an inexpensive yet fascinating addition to any zoo or animal attraction. There are nearly one hundred species, ranging in size from two inches to more than two feet. One giant species is said to feed on birds and mice.

Chameleons drink a great deal of water and this rather surprised me at first. I knew that most lizards drink some water, but the chameleons drank so much that they would actually come over and drink it from a glass, holding on to the edge

A chameleon feeding. As soon as it sees the food, it approaches within "shooting" distance of its six-inch-long tongue. Then it draws a bead on the food with its strange, telescopic eye. The tongue shoots out and engulfs the food, which is quickly brought back into the jaws. The food is then chewed with the strong teeth and swallowed. Favorite foods of the chameleon are grasshoppers, crickets, and flies, but it will eat any small insect.

of the glass with their strange hands while they drank. In the wild, they probably drink from the early morning dew that collects in droplets upon the leaves and stems of the trees. I have seen them doing this in my outside cage. But I also

Chameleons quickly become tame and will take food from your fingers. The food must be held lightly and released as soon as the chameleon touches it with its tongue. This is important, for the chameleon's tongue is very lightly attached to its mouth and can be easily pulled from the mouth should it stick to your fingers. This happened to one of the author's pets. The lizard survived without its tongue, but had to be hand fed. It later appeared to be growing a new tongue.

provided them with a large seashell of water in addition to a fine spray of water over the plants each day so that they could drink from the leaves.

Having kept all kinds of strange creatures in my lifetime, I would say that the old-world chameleon is probably the most fascinating of the animal world. It is easy to keep, simple to confine, makes no noise, is practically odorless, and is completely harmless. What more could be asked of a pet? It is not known how long they will live in captivity but I would say that a year or two would be the maximum, provided that they are not dehydrated and are in good health when

A grasshopper a split second before it enters the lizard's mouth. The chameleon would eat four to six grasshoppers a day. Note how the chameleon has changed its color to match the log. Some scientists state that chameleons change color only because of temperature change and not for camouflage, but the author definitely disagrees, having observed living specimens for a number of years. They can also change color instantly, although some books report that it takes ten minutes or more. They drink a great deal of water. The author has even had them drink from a hand-held glass.

acquired. They are tropical creatures and must be protected from the cold or they will die. A temperature of seventy to eighty degrees is ideal for them, and they should be kept in a place where they can sun themselves at various times of the day. They cannot stand cold or near-freezing weather. Even here in South Florida, I brought my chameleons inside during our cold spells in

A large old-world chameleon from Africa or Madagascar. The author kept a number of them in a large, outdoor cage that was planted with trees and shrubs. They made fascinating pets and quickly became tame, accepting food from his hands.

the winter months. I would gather them up just before sunset and place them in a large aquarium in the garage that was amply planted with small sticks and logs. Each chameleon must have a sturdy twig upon which to curl up and sleep. They will wrap their strong prehensile tail firmly about a branch or limb and go into a deep sleep as soon as the sun has set.

They are very sound sleepers, and it is best not to disturb them once they have fallen asleep because it seems to panic them if they are suddenly awakened. If you must move them inside during cool weather, move them while it is still light and they are awake so that they will have time to resettle themselves for the night in the indoor enclosure. If you pick up the chameleons when they are asleep it could harm them. They will open their mouths and clamp down on anything in reach, which may damage their jaws and tongue. They of course will clamp down on a finger if you stick it in their mouth and their jaws are powerful enough so that in extricating your finger you could easily break their teeth or jawbone. I have never been bitten by an old-world chameleon, but the larger ones probably could give a meaningful bite if you are foolish enough to stick your finger into their open mouth. They rarely if ever try to bite except at night when they are awakened from a sound sleep, and this is probably a purely reflex action. They are the most inoffensive of all lizards. Even a fast walk is beyond their capability. I have kept a fair number of them and have never seen one bite another, although they will open their mouth when another passes too closely. This seems to represent their maximum ferocity, unless of course you stick your finger in their mouth when it is open.

Some chameleons have live babies, others lay eggs, so that when keeping them, you may be blessed with a huge family of young chameleons if you are fortunate. One specimen that I kept gave birth to sixteen babies, all born alive and perfect duplications of their mother. They were about an inch long when born and immediately began crawling about their terrarium in search of food. It is difficult to raise the babies unless you have a large quantity of fruit flies or other small insects. Perhaps the easiest way to do it is to place them in a screen-mesh enclosure and push it down on top of some tall grass and weeds. There are usually many small insects living in the grass, and the young chameleons will feed on them. You can increase the food supply by placing a banana or a scrap of meat inside the enclosure, which will

Close-up of the three-horned chameleon shows long horns and strange, telescopic eyes that are on a swivel turret. Each eye can move independently of the other and can spot a small insect several feet away.

attract flies and insects. The enclosure should be of small-mesh wire, quarter-inch mesh or smaller, small enough so that the chameleons cannot crawl through but large enough to admit insects. Screen wire is too small, for it will keep out all of the insects that are needed as food. Take care that you do not place the enclosure near an anthill, because the ants will kill the chameleons.

The three-horned chameleon's prehensile tail is tightly wrapped about a branch, and its tiny hands and feet tightly grasp the limbs. At night the tail is rolled into a tight coil while the chameleon sleeps.

The large chameleons can be kept outside in the warmer climates except for cold nights in the winter months. Quarter-inch or half-inch mesh wire or hardware cloth is best for their cage because this will confine the lizards yet will allow flies and other insects to enter the enclosure as food. You will need to add extra food such as grasshoppers and crickets to be certain that your pets get enough to eat. Sometimes by placing a chunk of raw fish or meat inside the cage, enough flies will enter so that there will be plenty of food, but you will have to watch your pets to make certain that they are actually eating. If they do not get enough food on a regular basis then they certainly will not live very long. Also, do not neglect their drinking water. Their cage should have a rich, earth floor and be well planted with bushy trees and shrubs. If finances permit, it should be tall enough so that you can walk right inside and stand up. A pool at one end will enhance it and provide a constant source of drinking water.

In colder climates, an indoor cage will be needed. Again, the cage should be well planted with numerous branches, with a water dish or two on the floor. Screen wire is best for the inside cage, because you will want to confine insects to the cage so that the chameleons can eat them. If you are keeping the smaller chameleons, a fifty-gallon aquarium, converted to a terrarium, will make a nice home for them. In summer, you may be able to find many flies and grasshoppers, but when cold weather arrives, your source of supply will cease to exist. At this time, you may be able to purchase live food from your local pet shop or you may order live crickets through the mail from bait dealers. These are ideal food for the chameleons and are always available on a year-round basis since they are raised commercially by bait dealers. You can find bait dealers listed in any sporting magazine.

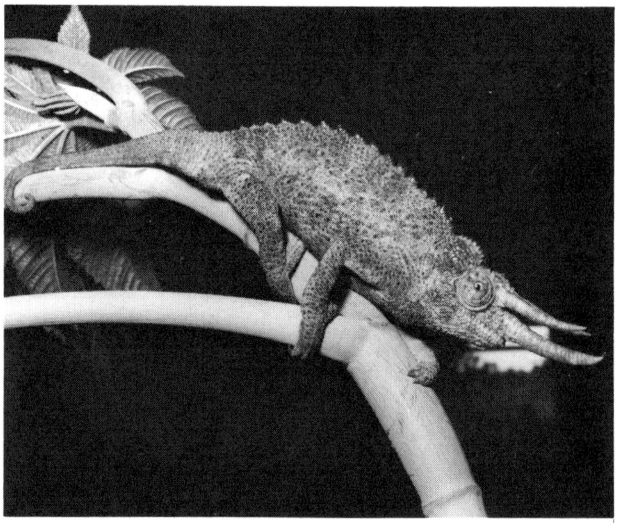

The three-horned chameleon makes an ideal pet, for not only is it intensely interesting, but it is harmless. It walks at a snail's pace, slowly moving one foot at a time. Its cage should be well planted with shrubs and branches and should have alternate shade and sunlit areas so that it can sun itself at will.

Chameleons are living wonders, and if you get one or several, make certain to feed them regularly and keep them warm. Be sure that you have a source of live food on hand before you order your pets, for they are sure to be hungry when you get them. I have never tried freezing grasshoppers and crickets in the months that they are plentiful but I suspect it would work. My chameleon would accept dead grasshoppers if I wriggled them a little, so it might be feasible to harvest live food for them in the summer when it is easily collected and have a full supply for the long winter months. I have never tried the large locusts that often plague countries in Africa by the trillions, but if they were acceptable as food for the chameleons as well as marmosets and other zoo specimens, it might prove to be a worthwhile business for someone in that country. Who knows? Someday someone may be exporting locusts from Africa as specialized pet food.

There are probably a hundred different species of old-world chameleons. All are fascinating. Some are tiny dwarfs, barely an inch long. At night they sleep on stems of grass instead of in trees. Others are giants, growing to nearly three feet. Instead of feeding on insects, they feed on mice and birds. I suspect that feeding one would be a rather expensive proposition. The average old-world chameleons are from five to eight inches long. They are fairly expensive. Good specimens may cost from eight to twenty dollars, but they are worth it for they are such unusual creatures. If you are fortunate in obtaining a pair or get a pregnant female who produces young, it would be a profitable venture if you could raise the young to maturity, especially if you could establish a breeding colony. You could make a major contribution to conservation, for like much of our wildlife, even certain species of chameleons may be on their way to extinction as man destroys and burns the forest for agriculture, or, even worse, sprays it with defoliants to kill the trees and brush, crops and people, in warfare.

Should you succeed in raising chameleons through succeeding generations, it could develop into a profitable business, for the supply of them is quite uncertain. With increasing regulations and downright banning on importing many species of animals, birds, and fish, it would not be surprising if lizards were eventually added to the long list of livestock that cannot be brought into our country. Raising them within our own country would be a welcome solution to this problem should it arise.

I would like to see at least some species of the old-world chameleon become established in the New World. Perhaps it could be done on some of our nearby tropical islands. Surely this slow, inoffensive creature that feeds exclusively on insects would never become a nuisance. It would never eat chickens or rob bird nests like the introduced mongoose or ravish grazing land like the rabbit. Even in the countries where it lives it is never really abundant. There is too much hogwash about introduced species, how they get out of hand and destroy the ecology of a country, but this is not true of all species. I have heard that the American horned toad is now living wild in the outskirts of Paris, but I have not heard of that country suffering a plague of them, nor is this likely ever to happen. The introduced ring-necked pheasant in our northern states certainly has not multiplied all out of proportion to other species, nor is it ever likely to do so. Neither would the little slow-motion chameleon. All he would do is eat bugs.

The new-world chameleon which is not a true chameleon is instead an anole (*Anolis carolinensis*), which is an iguanid lizard. It can change color, sometimes very rapidly, but it is not even distantly related to the old-world species and is almost the opposite in behavior and temperament. Instead of walking slowly, like the true chameleon, the anolis lizard runs extremely fast. It also bites if handled, but it is such a small lizard that the bite is of no consequence, amounting to nothing more than a pinch. Nor does it have a prehensile tail or the clasping hands of the old-world chameleon. However, despite the fact that it is not a true chameleon in the eyes of scientists, the general public invariably refers to it as *chameleon,* and this misnomer is likely to be attached to it as long as the little lizard exists. Care and feeding of the little anolis lizard is similar to that of the larger chameleon except that the anolis must be fed smaller food because of its size. Houseflies and baby grasshoppers are excellent food, and in winter it will eat small mealworms and beetles. It should be kept in a large, planted terrarium with green plants. The terrarium should be sprayed lightly each day to provide drinking water and should receive some sunlight if possible.

The Cuban anolis, also called the knight anole (*Anolis equestris*), is not as docile as the smaller American species. It grows to well over a foot in length and can inflict a painful bite. It is said to have become established in the South Florida area, but I have never seen any out in the wild. It is an interesting creature and closely resembles its smaller relative except that it is jumbo in size. Like the smaller anole, it runs very fast, and both

Paul Straughan examines a large flap-necked chameleon at close range.

The three-horned chameleon perches on a branch for Paul Straughan.

Paul Straughan holds an array of rare old-world chameleons. Despite their fierce appearance, they are totally harmless and walk in slow motion. The open mouth of the larger chameleon is a gesture of defiance or challenge to the three-horned and is the total extent of aggressiveness in most chameleons.

The tall pine tree in which the author's iguana lived. It would climb to the highest branch at the top of the tree to rest and sun itself.

species are good jumpers. Its cage must have a tight-fitting lid, and when the cage is opened for feeding it must be done quickly or the lizard will escape and be up the nearest tree before you can turn your head. The small anolis lizard is very inexpensive, usually selling for less than a dollar, or you can come to Florida and catch one free while it is running up a palm tree.

The green iguana is another interesting lizard that is easily obtainable (at present) and quite inexpensive. It looks very much like a real dragon, yet small specimens are completely harmless and quite gentle if handled with care. It is commonly available at most pet shops in sizes of ten to fifteen inches and makes a good pet for someone who wants something a little different. It grows to a length of six or seven feet, and a large specimen can inflict a painful bite, but the specimens usually offered for sale are mere babies.

My son, Paul, raised an iguana from a ten-inch baby to a fine specimen over two feet in length. He fed it chunks of banana, canned dog food, green leafy lettuce, and hibiscus flowers. It became very tame and would take food from his hands. But while we were on vacation, it somehow got out of its cage. We thought a skunk had killed it and we gave it up as dead. However, several months later, the iguana was sighted in

The iguana climbing down from the tree to feed on tender leaves near the ground. The tree was only ten feet from the author's house, and he could watch the interesting creature feed from his back porch.

one of the tallest trees in our yard. It was perched at the very top of a huge pine tree. We were thrilled at sighting it and noted that it had grown considerably since we had last seen it.

We watched it closely after that, and occasionally it would come down from the tree to feed on the bright flowers near the house. It lived outside in the trees for over a year and grew to nearly four feet, but eventually it disappeared. Perhaps it went away to seek a mate. There are said to be many green iguanas now living in South Florida, although some authorities state that the cold kills them or at least keeps them in check. But then, there are reports of all types of reptiles now living in our state—giant anacondas, boa constrictors, even piranhas. Someday, someone will report a wild elephant out there in the Everglades. It would not surprise me at all.

The rhino iguana, with its horns, huge bulk, and powerful claws, is practically a real, live dinosaur. It is a large, powerful lizard, four to five feet in length, with a heavy body and dangerous jaws. Although not quite as long as the green iguana, it is in a class of its own. It will actually charge you if you get it cornered and can inflict a very dangerous bite. I was intrigued with the prehistoric appearance of the huge lizard and ordered a pair from the famous animal dealer Bill Chase in Miami. I told him I wanted the largest specimens he could get and he said he would get them for me.

I constructed a huge cage for them in the backyard of strong, two-by-four welded wire, with a wire roof so that they could not climb out. Then I lined the entire cage with large rocks, which were placed against and on the wire to keep the lizards from digging out under the wire. The cage was about forty feet long, six feet high, and twenty feet wide. It contained small trees and shrubs with numerous large boulders to make the creatures feel completely at home. It was a nice cage and I was quite proud of it. I constructed a house of two large crates, placed one on top of the other, with openings so that the lizards would enter through a hole in front, then climb through another hole in the end of the crate to reach the upstairs level. I figured that this would give them a dark, secure home with complete privacy. It was covered with tar paper to waterproof it.

Finally my iguanas arrived and I was very excited. Bill had obtained two huge monsters for me, just what I wanted. They were male and female. They arrived in a heavy crate with strong slats nailed across the top so that they could get air. My daughter, Julie, looked through the slots and could see only a primitive eye looking back at her. She said that they had terrible faces and she was a little frightened of them. I kept the huge brutes in the garage overnight, for I wanted to release them into their new home in the daylight so that they could become accustomed to it and have plenty of time to settle down before nightfall.

The next morning, I moved the crate next to their new cage and carefully pried open the boards on top to allow the giant lizards to escape into the pen. As soon as they saw an opening in the box, they shot out into the pen and in seconds were at the wire on the far end, trying frantically to claw their way through it. I found that the two-by-four welded wire is not the best material for confining wild creatures, for soon the iguanas were tearing their faces on the sharp wire in their attempt to get away.

We quickly backed away from the cage so that they would calm down, and to my relief the huge lizards backed off the wire and began to explore their new home. It took very little time for them to find the crate that was to be their house, and as soon as they spotted the dark opening that marked the entrance, both scurried quickly inside. I could hear them climbing up into the top section, and I was pleased that they took to it so readily. They felt secure in the dark confines of the house and stayed in it most of the day. Only the male would venture out at all.

I found that the rhino iguanas had excellent eyesight, almost as good as a hawk, for they could see and recognize us from several hundred feet away. At the slightest movement on our part, they would quickly rush into the safety of the house. The male was much larger and heavier than the female, so it was easy to recognize, even at a distance. I had placed a huge dish of chopped apples, bananas, and other food in their cage, and I watched with anxious eyes to see if they would eat. But they paid no attention to it at first, because they were both very wild and unaccustomed to captivity. Actually, lizards and reptiles can go for a considerable length of time without eating, so I was a little premature in offering them food. Both specimens were extremely shy, and even the more aggressive male would not venture out into the cage for several days, except for very brief periods. But as they became accustomed to the surroundings, they stayed outside more and more.

The iguana caught in the act of actually eating a leaf from a tender vine in the author's yard. It would move from tree to tree and would let the author approach to within a few feet before it would scurry away at top speed.

This huge green iguana lived in the author's yard for more than a year. It had crawled out of its cage while the author was away on vacation, and he assumed that it was gone. But one day he discovered it in the highest branches of the pine tree. Paul Straughan was very excited because it was his personal pet, which he had carefully reared from a baby with large feedings of tender lettuce, flowers, and bananas. It grew to more than four feet long in the freedom of the yard but eventually went away, perhaps seeking a mate.

It was quite a thrill to look out and see my living dinosaurs sunning themselves on the large rocks that I had placed in their cage. The male came out by itself at first. I also saw it eating the food that I had placed in the cage. But the female was much more cautious. It was a full month after I got them that it took its first bite of food. Any movement in the yard would send it rushing back into the house. It looked thin and haggard, but eventually it came out to eat and sun itself with the male. It was a magnificent sight to watch them stretched out on rocks in the bright sun. It was like having a prehistoric world transplanted to my backyard.

Rhino iguanas are extremely hardy and adapt well to life in captivity. I was overly concerned because the female had not eaten for so long but then I learned that they could go for months

Close-up of a rhino iguana shows its prehistoric appearance.

without food. Lizards do not usually eat at all unless the weather is quite warm. In cold weather they apparently hibernate. I was well pleased with my huge specimens and had hopes that they would mate and reproduce, which was my main reason in obtaining them. Their cage was well constructed and had both rich loam and sand areas, as well as numerous rocks, grass, and shrubs so that they could find suitable cover for laying their eggs. However, they never seemed to settle down and were always trying to escape. Perhaps I had obtained specimens that were *too* wild. If I had taken some that had been in captivity for a year or two, I might have had better results.

The huge male would chew on the wire of the cage every time I approached, and as I looked at it, I would visualize a giant tyrannosaurus tearing at its prey with powerful jaws. My daughter, Julie, sort of liked them now that she could see them out in the open, but she was still afraid of them because of their fierce behavior. All went well with the iguanas and I looked forward to keeping them for a long while, until one day I

The rhino iguana is a powerful lizard and can inflict a severe bite. It grows to a length of about five feet. The author kept a pair of them in his backyard, attempting to raise them. They made interesting pets but never became tame. This fellow is shedding his skin.

was shocked to find one of them running loose in our backyard. It had somehow made a hole in its cage and escaped. It was the big male and it was headed for the outer fence. On the other side of the fence was a woodland of dense palmettos and brush, and if it ever got loose in that we would probably never find it. I ran in front of the iguana and chased it toward its cage. To my surprise and relief, it ran right back inside by itself. I looked at the section through which it had entered and found a tiny hole, hardly bigger than its body. It apparently had chewed the hole by constantly working on the wire with its powerful jaws. I patched up the hole firmly with new wire, but now that the rhino had tasted freedom, there was no stopping it. The next day, both the male and female were loose in the yard. Who would have thought that lizards were so intelligent? It took the male no time at all to communicate to its mate that it had found a way out.

I chased them back into the cage and patched up the new hole that they had made in the wire, taking great pains with it so that they could not break it open again. I replaced a whole section of wire so that there would be no weakened area in it, but I was very worried. I knew that once an animal escapes, it invariably will do so again. But apparently this time I foiled their escape, for weeks went by and they seemed to settle down. They would sun themselves on the rocks and move slowly and confidently about their cage with little of the panic they had shown when I first acquired them. I could approach them within a few feet now, especially the male, and could almost stroke it as it ate its food or sunned itself. I could not help but marvel at how closely they resembled the ancient dinosaurs. I had always planned to set up a "Lost World" exhibit, with authentic duplications of prehistoric scenes and mechanical, lifelike reproductions of the giant dinosaurs, using the smaller, present-day lizards to add actual living specimens to the display. I had even thought of selective breeding of such reptiles as the rhino iguana to increase the size of the family into a larger creature. Who knows? With special vitamin feeding and selective breeding of only the very

The author's children enjoying the iguanas in this natural setting. A cement-walled pit holds the reptiles. Natural trees and bushes grow up from the ground. This is an ideal way to keep and display reptiles. There are no bars or wire cage bottoms, and maintenance and cage cleaning are greatly reduced. Both mammals and reptiles will reproduce readily under such conditions. Crandon Park Zoo has already had several births of the strange aardvark, which is kept in an open-earth type of enclosure with a concrete wall, similar to the one shown above.

Green iguanas sunning themselves in the trees in their natural enclosure at Crandon Park Zoo in Miami.

largest specimens, perhaps someday one could raise rhino iguanas fifteen feet tall. Then we would have real, live dinosaurs. But alas, my rhino iguana project ran into difficulties.

One day I noticed that I could only see the female in the cage. At first I thought that perhaps the male was staying inside the house, but after a few days I realized that it had escaped—this time for good. I carefully examined the entire cage, and sure enough, there was a small hole chewed in the wire. It was a small opening and one would hardly expect such a huge lizard to pass through it, but I had seen the lizard do it twice before through a similar hole. Apparently it had escaped while its mate was still in the house. I was disappointed at the loss, for I had gone through so much effort to obtain and keep the mated pair. I had become very fond of them. Reluctantly, I sold the lone female, because I had no wish to keep it by itself. It would have been too sad.

I learned later that the big male had been captured and placed in a local tourist attraction. Apparently they thought it was one of their own specimens that had escaped. At this date, it is still living and in fine health (six years from the day it escaped). I have not given up on rhino iguanas completely. Someday I hope to try them again with a different type of enclosure. Like the green iguanas, rhinos are now said to be living out in the wild in South Florida. A number of young specimens were reported on Key Biscayne a few years back, but it is not known whether they survived hurricanes that struck the area.

Those who wish to keep these interesting creatures should realize that they are extremely powerful and are not easily confined. If welded wire is to be used for their cage, I would suggest one-half-inch-by-one-inch mesh wire. It should be buried in the ground or set in cement, and of course it must have a roof or the iguanas will climb right out. In addition, if the soil is soft, the wire will need to cover the entire bottom, because rhino iguanas are great diggers.

Probably the best way to keep them would be to build a smooth cement wall around the area selected for them. The wall should extend into the ground at least two feet, with a height of about five feet. Many reptile exhibits use this method for confining their snakes, and it works equally well for lizards. No top is needed, for the smooth walls offer no foothold and the lizards cannot climb it. The enclosure should be circular or oval rather than square so that the lizards will never come to the end when walking along the wall. This is a rather expensive pen, but it will last a lifetime with little maintenance other than occasional painting. The big cost is labor, so that if you do the work yourself, it will not be nearly as costly. It makes an excellent home for the lizards as well as for turtles, snakes, crocodiles, and numerous mammals. A round enclosure thirty to forty feet across would be ideal for most of the larger iguanas but, of course, a larger enclosure would be even better. Make certain to include a few small trees and shrubs in the center, some large rocks, and a natural pond.

The rhino iguana is an interesting creature. It is intelligent and alert. But it is also powerful and can be dangerous. It would not be a good pet for children. Rhinos are expensive. A good pair will cost close to a hundred dollars, more or less. If someone could succeed in raising them on a regular basis, he could probably make a fair profit. Of all the iguanas it is one of the hardiest but is also the meanest and toughest of the species. Still one cannot help but like them for their brute strength alone. Should you decide to get one or a pair, make certain that you have a secure home for them or they will escape as mine did. You may be able to tame them eventually and get them to eat from your hand, especially if you get specimens that have been in captivity for some time. But even if you do not tame them, you will still have one of the closest things to a real dinosaur, a living remembrance of the past when giant reptiles roamed the face of the earth.

Other strange reptiles that I have kept include the collared lizard (*Crotaphytus collaris*) from the western United States, and the chuckwalla (*Sauromalus obesus*), an interesting resident of the southwestern desert. Both are interesting creatures and look much like the dinosaurs of old. The collared lizard, which is a western species, is a beautiful lizard with handsome coloring. It is very active and alert. I had one that would run happily about the cage and come right up to me to take food from my fingers. It was my favorite lizard. It is not an expensive lizard and is easily confined with screen wire or hardware cloth. Like other lizards, it should be provided with both sunshine and shaded areas, as well as drinking water. It should also have large rocks and an open area in which to run. It makes a delightful pet.

The chuckwalla has a real dinosaur appearance. It is an interesting lizard but quite sluggish and apparently not very intelligent. It likes the hot sun but should also be provided with some shade. It is not an expensive reptile, and a good specimen can be obtained for less than ten dollars. There are many other lizards you can add to your dinosaur collection, such as the basilisk, the monitor, the sandfish, the moloch, the water lizard, and the gecko, just to mention a few. Then of course there are the famed tautara and komodo dragon, should you ever be fortunate enough to obtain either of these prized species. Of course it is always possible that if you had a natural, appropriate place for the lizards, the authorities just might release a few specimens into your custody as a conservation measure.

You can usually get specific instructions for keeping the more specialized lizards from the dealer who sells them to you. It is always best to do a little research beforehand into the feeding and habits of any species that you intend to acquire. There are a number of excellent books on lizards and reptiles in general. In addition, your local animal dealer may be able to answer questions that cannot be answered in any book at present, particularly questions on newly imported species. There is nothing quite like keeping a "living dinosaur" in your backyard or terrarium. With a little imagination, you can almost picture him roaming the earth millions of years ago when he was a thirty-foot monster.

6
Two-hundred-pound Rats

The capybara is the world's largest rodent. This strange but lovable creature comes from the swamps and rivers of South American jungles. It is truly the giant of the rodent family, for adult specimens weigh as much as two or even three hundred pounds with a body length of three to four feet. If the capybara had a tail like a rat, it would be seven to eight feet long, but it has no tail at all. It looks somewhat like a very large guinea pig.

My first capybara was an adorable, tame, affectionate, twenty-pound baby. Apparently it had been a pet of some Indian child in the jungles, for when I acquired it from Bill Chase, it was so tame that it thought it was one of the family. It would try to kiss everyone with its huge, bristled face. It was homely but with a personality and charm that was unparalleled to any wild animal I had ever seen. When I first brought it home, I turned it loose in the yard with the dog, cats, chickens, rabbits, kangaroos, and ostriches. But it was not content to stay outside with the animals. It followed me right inside the house. When I lay down on the sofa to rest, it crawled up on my lap and put its huge head against my neck. Then it immediately fell sound asleep. This would not have been so bad with a soft, cuddly kitten or even a soft-haired pup, but the capybara is a water animal like the beaver and muskrat. Its hair is very coarse and its face bristles with strong, stiff hair. To have a capy snuggle up to you is like being caressed with a stiff wire brush. I took it as long as I could, then I put the capy outside, but it let out a loud, whistling noise that could be heard all over the neighborhood and would not stop until I let it back inside. It would not stay off of my lap, so finally I decided to appeal to its stomach by pouring it a large glass of milk. While it was slopping up the milk, getting half of it on the floor, I sneaked away and went into the bedroom, quietly closing the door. But the capy did not like that at all, and soon I could hear it screeching and prowling about the house looking for me. It was certainly a demanding but lovable animal. I finally came out and petted it for a while, and it put its huge, fat head up against my foot and promptly fell asleep. If I got up to walk away, it would let out a loud cry. Finally I let the dog and cat into the house and all three of them curled up on the rug together and went to sleep. I always wish that I had taken a picture of them together. It was quite a sight.

I decided to get more capybaras and soon had three living in the yard. The new specimens were not tame, since they were straight from the jungles, but Mr. Chase, the animal dealer, had said that they would tame rapidly. He was right, for in no time at all I could stroke them. They eventually became so tame that they would push each other out of the way so that they could be petted. They would lie flat on their backs so that we could stroke their huge stomachs. One of the larger ones, who weighed about fifty pounds, would jump on my lap, nearly knocking me over, and nuzzle his huge whiskered face into my ear.

Like my other animals, I had acquired the capybaras with hopes of eventually raising them. I kept them in a large area in our backyard where they had good grazing area and provided them with a large bathtub, since they do require water. They would sit in the water and play in it much like an otter, then go running out into the yard, rolling in the grass to dry themselves off. I had heard that they are very intelligent animals, and

The author's tame capybara was one of his most lovable pets. Despite its ratlike appearance and bristling, beard-like whiskers, it is considered to be one of the most intelligent of all animals.

after having kept three of them, I would say that they are among the most intelligent of all animals, probably rating higher than a dog or a horse. Despite their strange, unorthodox appearance, they are surprisingly gentle. They quickly learned to recognize my voice and came running to me when I called.

They ate gigantic quantities of food and would eat just about anything. I had allowed them freedom of the entire yard and they quickly chewed down every small tree in the yard, including all of our palmetto trees, so finally I had to keep them in their pen lest they destroy the entire yard. One night they got out of their pen and the three of them attacked a ten-foot-high palmetto tree. They ate all of the branches, then chewed on the stump and ate it right down to the bare ground. We could hear them happily chewing on the tree all night long. They have huge, beaverlike teeth, two or three inches long, but my capys never bit me. In fact I could open their mouth with my hands to examine their teeth with little worry of being bitten. My first tame capy apparently helped to tame the others, for soon they were competing with him for affection.

Feeding the capys was expensive, for they ate all the time. They would eat an apple in two seconds and a banana would go down in three quick

The capybara can attain a weight of over two hundred pounds when fully grown. The author purchased this extremely tame animal from Bill Chase, famous animal dealer in Miami. The capy was so tame that it would follow the author into the house and jump up on his lap, demanding affection.

bites. I began to feed them hay, horse food, rabbit food, lettuce leaves, and sugar cane. There was just no way to fill them up. They soon outgrew the washtub that was their only swimming pool, and this was replaced with a full-sized bathtub, sunk into the ground. But they soon outgrew that and had to be content with it, since it was the largest swimming container I could get for them. They need a good-sized pond (with a good drain) if they are to be really contented, for they are strictly water animals. I found that mine were not getting enough shade or water because their skin became very rough and scaly, which did not help their appearance. I treated their skin with medication and it would get better for a time, but then

One of the capybaras, with the pitiful dry skin area about the face, probably caused by too much sunlight and not enough swimming area. The condition would disappear during the rainy season. It was most difficult to control.

The author's daughter, Julie, with a tame capybara. The animal was so tame that it would follow the family into the house and jump onto the couch in the living room.

the condition would return. I feel that this was caused from too much sunlight and insufficient swimming area, because during the rainy season the condition would disappear. Unfortunately, the capys had eaten all of the trees in their compound that were intended for shade.

Capybaras are charming animals. They are intelligent, gentle creatures, despite their fierce appearance, and can become as tame and affectionate as the family dog. But they need a large enclosure

The capybaras taking a bath in a large washtub. They loved the water and spent most of their time in it, which is natural since they are an aquatic species.

with a strong, six-foot fence, buried partway into the ground. They are powerful creatures and can push their way through a weak fence with little effort. I lost my capys this way. They somehow got out through the fence. One got out by pushing the gate open; the other went through a hole under the fence that had been dug by one of the neighbor's dogs. The third capy died during very cold weather. It escaped from its warm house and froze to death out in the open yard during the night. I was saddened by the loss, for I had developed a genuine fondness for the animals. They had been faithful friends and I felt that I had somehow let them down.

I decided not to get any more until I could provide a good habitat for them. I feel that they should have a grazing area of five to ten acres, with a pond of at least an acre, preferably with running water, as they tend to dirty the water rapidly. In cold weather they must be protected or they will surely freeze, so either the pond would need to be heated or they would need a heated house in the winter months, even in Florida. Up north, they would need to be protected from the cold all year round, except for the warmest days in summer.

The capybara is a valuable zoo animal. Full-grown animals are worth several hundred dollars.

After a swim, the capys would roll on the ground and frolic in the sun.

Young capys cost thirty to fifty dollars each depending on their size and weight. I do not know of anyone who has had them breed in captivity, but it would certainly be worth a try. Perhaps some of the zoos have succeeded in breeding them, since some of the zoos are now keeping them in good-sized paddocks. If you decide to try keeping them, perhaps my experience will be of value to you. There is still a lot to be learned about them. I heard later that one of my animals that had escaped was shot by a hunter in one of the canals near by, a sad fate for such a noble animal. The hunter probably thought he was shooting at a giant rat. Some people are so quick to kill anything that moves or swims. I intend to try a colony of capys again sometime in the future, but when I do, I will try to do it right and make their habitat almost a duplicate of their jungle home. This way, perhaps, I can raise them to the full two- or three-hundred-pound size, which would be quite a remarkable animal. After all, how many people have ever seen a two-hundred-pound rodent?

The capybaras sunning themselves in the yard. They were voracious eaters and soon ate every small tree and bush, destroying their shade, which resulted in severe drying of the skin. These animals should have a good-sized pond to be content, yet the author has seen them in some public zoos with no water whatsoever.

7
The Best Pets of All

Ocelots are intriguing, baby lions are really cute, honey bears are fascinating, as are chimps, mountain lions, and baby polar bears. But all of these will present you with numerous problems, particularly when they mature. However, there *is* an animal that will bring delight and endless joy to your children and yourself, yet is inexpensive and safe, and keeping it will present practically no problems. This is the common house cat, a proven pet, and this is a fact.

The reasons are many. First of all, the house cat has been domesticated for hundreds of years. It *likes* living with humans. If you treat it with kindness (as you should) you will have to drive it out of the house, for it would rather stay in the house with you than remain outside. In contrast to a wild animal, which is constantly seeking to escape, the house cat is constantly seeking human companionship. Is this not the whole idea of having a pet?

We have had numerous cats in our household. I will have to admit that I have not been overenthused with our Mother Cat, who has faithfully blessed us with a batch of kittens at least twice a year for the past five years. But I do know that the kittens she has produced with such aplomb have brought our children endless joy. My daughter, Julie, would play with the little kittens for hours on end, marveling at their adorably cute faces and carrying them about the yard in a cardboard box or gently nestled in a group in her arms. The baby kittens have taught her a love and respect for animals that could not be learned in school or from any book. They have taught her kindness and tenderness, as I am sure they have taught other children.

My son, Paul, also has enjoyed the kittens, though not quite as much, since boys have so many other interests and activities. But he would play with them when he was small. He would build little houses for them of cardboard with numerous doors and openings, and watch them climb from room to room as they chased each other's tails and playfully fought each other for hours on end. Paul took a liking to an orange tabby, and since it was a male, we decided to keep it. It grew into a huge tomcat and was named Diaper Rash, a name that the children had picked up from a television commercial. Diaper became a true friend and was never hostile in any way to our son. The huge cat would purr very loudly and would let Paul carry it about the yard with its long legs often dragging the ground, since it was nearly as tall as my son when it stretched out. It grew up with our son in his younger years, and many nights we would find the huge tomcat snuggled up with Paul on the bed, both sound asleep. Diaper was with us a little over two years. One time the end of his tail was bitten off, probably by a neighbor's dog, and Paul accompanied me and his cat to the veterinarian, where the cat was treated and duly registered as *Diaper Rash,* which brought a few giggles from the receptionist. Diaper taught my son a kindness and respect for animals, as the little kittens had taught Julie, and the entire family enjoyed the big tomcat that was just a great big overgrown bundle of fur and affection.

Being a tomcat, Diaper would disappear from time to time, as do many male cats, and one day he did not return. My son missed him and his eyes would form tears when he thought of that huge orange head cuddled up on his bed. We felt that perhaps Diaper would eventually return but we never saw him again. He may have been struck

One time Mother Cat had her kittens in the woods. The author cautiously followed it and saw it disappear into a small hole in the ground, where he found the newborn kittens in a warm, snug nest. Mother Cat had hollowed out a deep hole beneath a pile of dead leaves and branches and had camouflaged it so carefully that the author walked by it for two weeks before finally discovering it.

by a car or found a new home, but eventually we gave him up as lost. We still had Mother Cat though, and she was producing as well as ever. We had nicknamed her *Fruitful* because she produced so many kittens. In one batch was another orange tabby that bore a close resemblance to Diaper Rash. It was also a male. My son announced that he would like to start another Diaper Rash, so we kept the little orange kitten. It soon grew up into a fine cat with all the gentleness and companionship of the original Diaper. It would run into the house as soon as we awakened in the morning and would settle down on the sofa. At night, it would quietly walk into my son's room, where it would curl up on his bed and fall asleep with him. It brought him much joy and happiness. But alas, it was eventually struck by a car and killed (cars kill so many pets and animals each year). Paul was saddened by the loss, and when Mother Cat presented us with the next batch of kittens, Julie excitedly announced that there was another orange tabby in the lot, and it was a male. Paul started a *third* Diaper Rash. This one didn't last very long. It was accidentally carried away in one of the family cars and disappeared. At this writing, Paul is now working on Diaper number four.

Few pets have the intelligence, charm, aloof-

After the secret hiding place had been discovered, Mother Cat moved the kittens nearer the house, this time hiding them inside a column of concrete blocks. Eventually the kittens ended up in a cardboard box on the back porch, which was the usual birthplace for Mother Cat's semi-annual batch of kittens.

Julie Straughan lovingly displays one of her kittens. A little kitten can bring much joy to a child and is a safe, dependable pet.

Mother Cat with its latest brood. What a pity that more people do not adopt kittens instead of exotic animals for their children. Kittens make much better pets, are inexpensive to obtain and keep, and are far less likely to cause injury to the child than a wild creature would.

ness, loyalty, and affection of the common house cat, with the possible exception of the dog. The cat is a practical animal. It is easily housebroken. It is extremely inexpensive to purchase or obtain. Usually, you can get one for nothing. It is a safe animal and is not likely to attack its owner as wild pets invariably do. Consider the kitten next time you decide to get a pet for your child or yourself. With a kitten you can hardly go wrong. It will add joy to the household and grow up as one of the family. Unlike the exotic animals, which admittedly are irresistibly cute, the common house cat will not bring serious, often insurmountable problems to you as it matures. It will just grow up and quickly adapt to your home. It is a sure thing.

The same cannot be said of any exotic animal. When it grows up, it is certain to present problems. If you *do not* like cats, there is always a puppy to consider as a pet. It too is a charming little animal, completely domesticated and accustomed to living with people. It too is a sure thing. I need not go into the devotion of man's best friend, for anyone who has kept a dog will realize that here is an animal that gives everything, for companionship and faithfulness to its owner and master. A tiny puppy is certainly as cute as any exotic animal, particularly a snow white baby huskie or a toy pomeranian. *All* puppies are cute and make wonderful pets. When they mature, they do not create problems like exotic animals do; instead, they become an asset to the household. They become a faithful and trusted companion to all members of the family, and in addition many

When it comes to pets, there is no animal more suitable than the common house cat. The little animals are domesticated and love humans. Unlike the exotic cats, which are constantly trying to get away, the house cat will constantly try to get into your house to be near you. Julie Straughan holds Spunky, which she has dressed for the occasion.

Little Spunky, one of the many cats in the author's household.

A baby Abyssinian cavy, just two days old. The young are born fully developed with wide-open eyes and a heavy coat of fur. The average litter will bring four to eight of the living "powder puffs."

The author raised hundreds of Abyssinian cavies (a domesticated type of Guinea pig) in this large, outdoor cage in Miami. Waterproof crates provided protection from rain.

The Abyssinian cavy, an ideal pet, especially for children. Unlike hamsters and white mice, cavies rarely bite. In fact, the author has raised hundreds and has never had one bite. They are inexpensive, highly intelligent, and very cuddly with their soft, curly fur. They can even whistle.

of them become watchdogs, barking a warning at any intruder. This alone more than justifies keeping them in these days of ever-increasing crime. You will not find many exotic pets that will warn you of a burglar or protect your property, unless you could train a full-grown gorilla to apprehend intruders. One encounter with a gorilla in your house will put it off limits to burglars for life. A dog is much more practical. There are so many different kinds of dogs to choose from. They range in size from two-pound animals to two-hundred-pound giants, and the most fastidious householder can find one that will exactly fit his or her needs. The common mongrel makes as good a pet as any, but for those with profit in mind, the pedigree animals should be selected, because they still make a good pet and the young can often be sold at a good profit when there is a market for them.

Those who want a different animal from the dog or cat should consider the guinea pig, or cavy, as it is known in the pet trade. I consider the Abyssinian cavy (a specialized breed of cavy with tight, curly hair) to be just about the ideal pet for a child. Having kept all types of small animals, including flying squirrels, hamsters, weasels, marmosets, skunks, woodchucks, and chipmunks, I can positively state that none of these animals can compare with the beautiful, gentle Abyssinian cavy. Unlike squirrels, chipmunks, ger-

Cavies are easy to keep and raise. They come in all colors, from jet black to orange to tricolor to snow white. One time a family of small, striped skunks moved in with the author's cavies, and believe it or not, a guinea pig was born with black and white stripes identical to those of a skunk.

bils, and hamsters, which are all very likely to take a bite out of you or your child, the little cavy will seldom, if ever, bite. I have raised and handled hundreds of them and have never had one bite me. Unless of course you stick your finger into its mouth, you are very unlikely to ever be bitten by the gentle cavy. But this does not hold true with the other animals. I have kept practically all of the most common animals in my various pet shops and have suffered a good many bites from chipmunks, white mice, white rats, golden mice, and hamsters as well as from most of the larger animals. The guinea pig or cavy is the exception among rodents for gentleness.

Small Abyssinians look like living powder puffs. They are a clean, practically odor-free animal and come in all colors, from snow white to black, tan, orange, or tricolor. I once had one that was black and white striped with the identical coloring of a skunk. A small Florida skunk or civet cat had moved in with my Abyssinians and slept with them for several weeks. It did not seem to be harming them in any way so I left it alone. One day I noted that one of the female cavies was expecting, and when the blessed event occurred, one of the guinea pigs was born with the exact coloring of a skunk. It had a little white line over each eye and the black and white striping down the back. It was incredible. I called Bill Chase about it, but he was sick and could not come over to see it. I decided to photograph it right away, but when I went out to the pen, much to my dismay the mother had killed it (which cavies and other animals sometimes do). I was told that such a mating is impossible, yet I had held the skunk-striped guinea pig in my hands. The skunk still sleeps with the animals, so I am hoping for another "sprinea skink" or whatever you would call a half skunk and half guinea pig!

Janet Gibby and her squirrel monkey, B.C. Her pup, Crash, cannot even be petted without B.C.'s becoming jealous. Courtesy Dr. John F. Gibby.

Other domestic animals that make fine pets are rabbits, ducks, and chickens. I have had many pet ducks and they can become a most affectionate pet. They will snuggle up to you and good-naturedly nip you with their strange bill, with true affection. Geese also make good pets and excellent watchdogs, but they require more living space because of their size. They really belong on a farm out in the country. They can also be dangerous at times if allowed to get out of hand.

The little bantam hens and roosters make interesting pets for those who live out in the country. City folks should stay clear of them, particularly the rooster, for many people would report the keeping of a rooster within city limits quicker than they would a criminal activity. I once gave a bantam hen and a rooster to a lady in Miami who wanted them for pets. She had them for less than a week when she was reported to the police by her neighbors. Police converged on her property a dozen times in less than a month, trying to catch her feeding the illegal birds so that they could establish ownership and arrest her. She said she wished that the police were as vigilant in combating crimes as they were in trying to arrest her for keeping two little chickens (such a ghastly crime). For country dwellers, the little bantams can make amusing backyard pets that will require little care and are full of surprises. They hide their eggs and lay them everywhere. We had one come into the house and lay an egg on top of the phone directory. Several times they laid eggs under the front seat of my auto when I left the window open. To watch a pair of bantams cavort about the yard is always amusing. It could almost be a lesson in human behavior. The rooster is a true gentleman, always the boss, but courteous to his lady. He will call to her when he finds a choice tidbit and lay it down in front of her to eat, rather than eat it himself. He will protect her from danger, but he expects her to be an obedient wife. She respects his strength, wisdom, and authority and follows him about the yard as a true wife should.

There is nothing quite as cute as baby chicks

Janet Gibby and B.C. The family had to get rid of B.C. because the zoning department of Coral Gables, Florida, said that monkeys are obscene and must be kept out of public view. Courtesy Dr. John F. Gibby.

running about the yard, peeping to their mother as it watches over them and teaches them about the rules of life. If you have a pair of bantams, sooner or later you may be enriched with a large addition to your chicken ranch. We started out with just one pair and soon we were inundated with peeping chickens and crowing roosters. Bantams are great sitters and will sit on a nest of eggs through rain and storm. Many farmers use them to hatch out the eggs of other birds, because they are better than an incubator. One time I put a huge ostrich egg under a sitting bantam as a joke, never expecting her to stay on it, but she stayed, even though it lifted her high into the air. Another time, we were missing one of our kittens and I searched throughout the yard for it, in vain. I finally came across a bantam hen sitting and noticed a small patch of black protruding from beneath her. Surely it could not be the kitten, I thought. I quietly moved the hen aside and sure enough, she was sitting on the kitten. The kitten was curled up snugly with an egg, purring loudly and completely contented. The little bantam did not mind it at all. Perhaps this was her first try at hatching eggs and she thought the kitten had hatched from an egg, not yet knowing what a baby chick looks like.

Even pigs make a good pet. They are considered among the most intelligent of animals. But again, make certain you check with local zoning laws before you acquire one (or any animal for that matter). One day I received a call from Dr. John Gibby, our children's doctor. He wanted to know where he could purchase a pig. I told him to look in the want ads, under *livestock,* in the newspaper, and did not give the matter much thought until months later when I got another call. It seems that Dr. Gibby got the pig all right,

A baby pig makes a cute pet, but remember that it will grow fast. This little fellow is the personal pet of Mary Prince.

Mary Prince and her pet pig.

but also a batch of turkeys, monkeys, and all kinds of pets. It was the pig that nearly proved his undoing. It had escaped from the yard and apparently aroused considerable attention in the extremely strictly zoned city of Coral Gables. In attempting to apprehend the pig, the turkeys and all the other pets of the family were brought to the attention of the zoning department. Dr. Gibby said that he had been arrested and had to go to court. The local newspapers made hay of it and the doctor was ordered to get rid of a number of his beloved pets. What a pity that zoning authorities cannot have a little more heart.

This brings up a good point. If you are moving out to the country and wish to keep animals, check on the zoning of the land in this respect *before* you purchase it. In most states, if you have land with agricultural zoning, you can keep all the animals you like, within reason, with no zoning worries. There are some people who hate animals and they will quickly report you to the police or zoning authorities, but if you have the law on your side, you will have nothing to worry about. The laws concerning exotic animals are often extreme and all-inclusive. There is always that "special permit" (which I abhor) on the books in most states, and as soon as it is discovered that you have an animal other than a cat or dog in your house, you are likely to be fined or arrested, unless you possess the special permit. A feature story in the newspaper about a local person with a pet lion or tiger makes interesting reading, but it is a sure bet that the local zoning officials will read the story with only prosecution in mind. I have seen this happen so many times. A person will get a little fame in the papers and a few weeks later he will be in deep trouble with zoning or local conservation agents who will try their best to take away his joy. When it comes to what is the law in our "Land of the Free," it would take a volume of books that would reach from here to the moon. Laws governing animals and wildlife are very complex, usually unsound and ridiculous, and they get worse each year. Pretty soon you will not be able to keep a flea on your dog's back without a special license for the flea, in addition to the license for the dog.

Thus, not only does the kitten, dog, or guinea pig make a more practical pet for the average household, from a legal standpoint there are bound to be fewer complications with the common domesticated animals than with exotic animals.

Janet Gibby with her seven-month-old friend, Oink. Oink is a little too large for the lap now—it must weigh 150 pounds. Oink likes dog food, but there is not enough in the world to satiate this animal. Courtesy Dr. John F. Gibby.

The inoffensive tortoise makes an ideal backyard pet. It is quiet, harmless, and extremely hardy. This fine specimen of the Berlandia tortoise belongs to Fred Leonard of Miami.

I am not stating that a dog or cat is more interesting than an exotic animal, but merely that it is more practical and is much easier to keep. Wild animals enjoy living out in the wild. The squirrels and rabbits love to roam about the countryside. They would not be happy in a cage. But the domesticated rabbit or guinea pig will not mind his cage at all, so long as you feed it and give it water to drink. Nor will your cat or dog mind sharing your home with you. They *like* to live with humans. Think about it when you decide to get a pet.

On the other hand, if you must have an exotic animal, talk with people in your town who have one or have kept one in the past. They may enlighten you to the problems and heartache that

Underside shot of the Berlandia tortoise showing interesting armor, powerful claws, and protective neck extension of the shell.

A beautiful South American tortoise (Testudo denticulata) *living outdoors in Fred Leonard's yard.*

The handsome yellow-footed tortoise from Ecuador also makes a fine pet. It grows to well over a foot in length and is long lived. This is also one of Fred Leonard's pets.

The red-footed tortoise from Peru is another favorite of turtle lovers. It is a colorful species with beautiful markings on the shell. Fred Leonard keeps all of his turtles in his yard confined with regular chain-link fence. Turtles are a favorite pet of many people throughout the world. There is even a turtle society and magazine especially devoted to these interesting creatures.

The Gulf Coast box turtle living in Fred Leonard's yard. The turtles were so contented in their protective haven that they laid eggs and produced young. This is one of the great joys of keeping any wild creature in protective custody, for if it reproduces, you can release the extras back into the wild, thereby assisting our rapidly diminishing wildlife.

Baby box turtles in their egg cases. The little fellows have just hatched from eggs laid in a sandy area in Fred Leonard's yard. Turtles make fine backyard pets. They are completely quiet and harmless.

A box turtle with newly hatched young.

often result in keeping a strange animal from another land in an environment completely alien to its original home. Should you decide to go ahead and get that ocelot or monkey anyhow, do yourself and the animal a favor by first getting a book or pamphlet on its care and feeding before you acquire it. Second, buy it from a reputable dealer and make certain that it is healthy and has its needed shots, in the case of a cat. Third, have its cage ready for it before you bring it home.

Another ideal backyard pet that will present very few problems is the tortoise. It is silent and inoffensive and may easily be confined with most any kind of wire-mesh fencing. Some species dig, so it may be necessary to run a portion of the wire underground for at least two or three feet. Tortoises do well outside, especially in the warmer parts of the country, and will even lay eggs and reproduce their species. Fred Leonard of Miami, who has kept a number of tortoises in his yard, reports that they have grown tremendously in size. He keeps some of the larger South American species as well as local types. Some have already reproduced, laying eggs that hatched out into cute little tortoises, a most worthwhile accomplishment for Mr. Leonard in providing proper care for his pets.

Four baby box turtles born and reared in Fred Leonard's yard.

Many of the larger tortoises, particularly the species of Aldabra tortoise from the Seychelles, which reaches a length of over four feet, and the Galapagos tortoise of similar dimensions, are in danger of extinction, should a change of government decide to take over their land for human habitation. Such giants of the turtle world would be worthwhile as a pet if one could obtain a mated pair with prospects of breeding them. They would need a good-sized yard with suitable area for laying their eggs. Unfortunately, some zoos keep them in tiny quarters much too insignificant for such a valuable species. A grassy backyard of an acre, with rocks and a few trees or brush for shade, would make them an ideal home. Of course they would need heated quarters for a northern climate.

There are many other interesting species of turtles or tortoises that would be suitable for a pet or as a prospect for breeding of a threatened species. Some are spectacular in beauty, such as the star tortoise (*Geochelone elegans*) from Madagascar. They are also quite valuable, and raising them would be well worthwhile. As with other pets, it is best to learn as much as possible about the specific turtle or tortoise that you intend to acquire. In addition to books on turtles, there is now a very fine magazine devoted entirely to the keeping of turtles and tortoises.* It gives detailed instructions for the feeding and raising of these interesting reptiles.

Perhaps in future years, we will see a great many of the exotic animals raised in captivity so that a person desiring a monkey or ocelot for a pet, instead of the conventional cat or rabbit, will

One of the baby box turtles, held in the hand to show its small size.

be able to obtain a semi-domesticated animal, which of course would make a much nicer pet than an animal from the wild. It would be a safer, disease-free, acclimated animal, and keeping it would be much less of a problem. Already, skunks are raised in quantity for pets and are a much more suitable animal than the wild, caught specimens that sometimes have rabies. All kinds of exotic animals could and should be raised for the prospective pet owner rather than presenting him with an animal straight from the wild, which is what is being done at present.

* *International Turtle and Tortoise Society Journal,* Box 45555, Los Angeles, California 90045.

8
Our Wildlife Is Dying

It is a fact that many of our most interesting wild animals are threatened with extinction, even in the most remote areas of the globe. Man has killed off wildlife at an unprecedented rate. Nowhere are wild animals safe in the wilds except in a well-patrolled national park or wildlife preserve. Even the giant whales face extinction in the frozen Arctic and Antarctic. They are no match for modern killer ships. Also threatened is the magnificent polar bear, who is hunted by plane in a white wilderness where it cannot possibly escape. Many of our most splendid animals have been reduced to mere hundreds as their total population in the entire world. Man alone is to blame.

Did you know that in Africa, one man killed nearly a thousand rhinoceros in recent years? It happened in Kenya and the man was paid to do it by the *government.* The animals were killed to make room for people. Other animals have met a similar fate. In all of Asia, the cheetah, once common throughout the land, is now practically extinct. The number of Indian rhino, one of the most spectacular beasts of the world, is now down to six or seven hundred animals. The Sumatran rhino is in even worse shape. Its numbers are down to a little over a hundred. The Javan rhino is in a critical state, for it is said that less than thirty-five animals now remain.

Many animals are on the endangered species list, and the number of threatened animals is growing each year as man kills the animals or destroys their habitat. Progress is slow in saving them, for even in civilized America a bounty is still paid for killing wolves and other animals. On the national bison range, excess bison are still being shot. What a fate for such a splendid creature. Why doesn't the government ship the excess animals (if they really are excess) to other national parks and public lands so that they can increase to at least a half million? The government pays farmers *not* to grow crops every year. Why not release bison on the farmlands that are not to be planted and let them roam the plains once more? We have few enough animals in the wild without killing off a so-called surplus every year.

Here in Florida, it is reported that there are fewer than fifty bears in the entire state. Yet hunting bears is still allowed. What kind of conservation is that? Do the officials want *all* of our wildlife to be ultimately destroyed? It appears that they do. Our own government kills off many of our animals directly and indirectly. It kills the bears, the bison, the wolves, and the coyotes. It even pays trappers to go off into remote areas to kill what little wildlife we have left. But worst of all, it allows the spraying of aerial pesticides, many of which have had a disastrous effect upon wildlife, sometimes making nearly a one hundred percent kill of birds and small mammals in the area.

Our wildlife is in imminent danger, and those who care about it should make their feelings known. Do not think that writing letters to government officials is a waste of time. It is not. I have won many small victories as a direct result of writing letters to the proper agencies. To whom should you write? In matters regarding wildlife or the despoliation of a wildlife area, you can start by writing to the head of your state conservation department, but do not let the matter end there or it *will* be a waste of time. Insert

A typical desert scene in Southern Arizona. The author has traveled considerably throughout the country and is appalled at the scarcity of wildlife, even in remote areas, such as the one shown above. The author walked through this area by himself on foot and saw no sign of game. On an eight-thousand-mile auto trip from Florida to California, visiting and staying at such national parks as Grand Canyon, Sequoia, and other wild areas, he saw practically no game except for a few squirrels and deer. In Africa's national parks, thousands of head of game may be seen. What has happened to our wildlife?

several extra sheets of paper with carbons into your typewriter and make copies of your letter. This is most important. Then send a copy of your letter to the governor, state senator, local newspaper editor, local television newscaster, and Fish and Wildlife Service in Washington, D. C. At the bottom of your letter make a note that copies were sent to these agencies. This way your letter will not be ignored. Do not be concerned about the proper department to which your letters are sent, for they will be forwarded to the correct department by the office that receives them. This type of letter writing *will* bring results, or at least a reply. A single letter usually will not. The forty or fifty cents you spend for the extra stamps will be money well invested, if you have a genuine interest in saving our wildlife.

You should get others to write, as well, particularly if you belong to a club or a society. Naturally the more letters on a subject, the better the results, for any vote-conscious governing body will be quick to please their constituents, regardless of the urgency of the cause. If you feel strongly about open hunting season on animals that are extremely rare in your state, by all means write. Otherwise little or nothing will be done. If you think hunting should not be allowed in state or national parks, *write!* If there is a beautiful wooded area or wilderness area near you and you think that it should be preserved in its natural state rather than allowed to be bulldozed into housing lots, write to your governor and local newspapers. Photos are of a great help in aiding your cause.

The wildlife in the United States is in a critical state. Not only has hunting decimated it to the verge of extinction in the past hundred years but, even worse, the habitat of the wildlife is being destroyed on an unprecedented scale. Forests are mowed down for highways, swamps are drained or flooded. Either way the wildlife is destroyed, for without its habitat it cannot survive. We have too many people, for one thing. Another is that the trend is to move farther out from the city, eating up all of the forest and field. The hardened city dweller may feel that the wildlife is of no importance and that only man should be considered when it comes to using the environment. But this is an error. If there were no birds, the world would swarm with insects. If there were no snakes, the woods would be taken over by rodents. Coyotes, hawks, owls, skunks—all help to keep rodents and insects in check. The trees and shrubs, that overgrown "jungle" lot in your neighborhood that local residents demand be cleared because a mugger or other criminal might hide there, must be viewed with a new light. Instead of clearing the lot down to the bare earth and billing the owner for the job, as is done in Miami and in many other cities, why not erect an eight-foot fence around the area, with no gate? This would protect the local people from any would-be bandits who would lurk there. It would also keep the area wild and provide a haven for birds and small mammals. More important, the oxygen produced by all the shrubbery would freshen the area. The trees would also absorb a great deal of dust and noise that would otherwise go directly into the homes of nearby residents. Why not consider this at the next meeting of your women's or sportsmen's club? Large lots could even be stocked with game that would be controlled and removed should the property eventually be sold or developed. You can do a great deal, as a club or a society by making such ideas into a project and working together to raise funds or to accomplish legislature to be passed, protecting our wildlife.

It is evident by the total pollution that is sweeping our nation that the state and local conservation departments have not been very effective in conserving our wildlife. In fact some conservation departments have changed their name to *Department of Natural Resources* instead of *Department of Conservation.* Apparently they realized that the former name would be easier to live with, for in fact they really did not conserve much, other than their jobs. It appears that conservation interests come second to big money corporations that would destroy the natural resources of an entire state and never be stopped or even hindered by the conservation department. In Florida, particularly South Florida, practically every river and fresh-water canal has been so thoroughly polluted that few if any fish or wildlife remain. The salt-water marshes and estuarine areas are in danger of total ruination by pollution, dredging, and development. Giant lakes are dying or dead, along with all the fish and wildlife in them, yet we still find conservation agents measuring bass and other fish. Instead, they should be measuring pollution. Then they would be *really* helping to save our wildlife.

Even giant coral reefs are now being killed off of the Florida coast. I was among the first to bring this startling news to the public view. Hundreds of thousands of tons of living coral has been killed by dredging down in the Florida Keys, and until I reported it in my *Salt Water Aquarium* magazine, this horrible destruction of our environment had not been reported anywhere. It is not that the conservation department did not know about it, for I even pointed it out to conservation agents who had approached me while I was in the area diving. But why was there no public outcry? Must every living thing in our country be killed before the people will act? Dead things cannot be brought back to life. If we allow our precious coral reefs, lakes, and forests and all of our wildlife to be destroyed by water and air pollution and land development, they will be lost forever. Already a number of birds and animals have been killed to extinction by man, and many are on the endangered species list. You can help save our wildlife by taking an active part in their preservation. Join a club, or start one, with the object of protecting our wildlife and environment. Start a chlorophyll club, whose purpose is to plant trees in barren, denuded areas. Or start a wildlife preservation club. Then lease, with an option to buy, large wildlife tracts, to keep them from the hands of the developer. You can make the area into a park with wooded trails and even name it after your club. Also, join the already established

A fearsome looking musk-ox bull at Alberta Game Farm. The animals are kept outdoors in a large, natural area similar to their native terrain. Courtesy A. F. Oeming, Alberta Game Farm, Alberta, Canada.

A baby camel born at the Alberta Game Farm. Note the beautiful, natural paddocks in which the animals are free to roam. Courtesy A. F. Oeming, Alberta Game Farm, Alberta, Canada.

nature groups like the National Audubon Society, National Wildlife Federation, World Wildlife Fund, Save the Redwood League, National Parks Association, Sierra Club, and other groups dedicated to saving our precious wilderness and wildlife. You can send them an annual (tax deductible) contribution each year, and it is money well spent. You can even leave them a sizable contribution in your will. Give at least some of your money to an organization that will carry on your own ideas of preservation of our wildlife.

The National Audubon Society, for example, actively preserves our wildlife with its sanctuaries, some which encompass thousands of acres. They purchase the land, patrol it, and keep it in its wild and beautiful state. They also publish a beautiful magazine and bring to the attention of the reader, as well as to the general public, all current or proposed government and private projects that will damage our environment. Is this not a worthy cause? You can join them for ten dollars per year. Write to: Audubon, 1130 Fifth Avenue, New York, New York 10028.

The World Wildlife Fund in Washington, D.C. is another active organization that operates on a worldwide scale to preserve our dwindling wildlife. Contributions to them could not be better spent. They actually buy large tracts of undeveloped land to provide a habitat for the animals. They assist foreign governments in the protection of game from poachers in the national parks of Africa, India, and other countries. They also fund research for better animal management (particularly of animals on the endangered species list), actively participating in the purchase of the land, which is the key to survival of all animals. This is a most worthwhile organization. Write to them at: World Wildlife Fund, 910 17th Street, N.W., Washington, D.C. 20006.

They need your help as do the other nature conservation organizations. They will keep you advised of wildlife that is in danger and will tell you how you can help to protect it, such as by the campaign to protect the spotted cats like the leopard, cheetah, jaguar, ocelot, and others. Many women do not seem to realize that these animals are in danger of extinction. The fur coats that they buy, made from the pelts of these animals, is bringing about the rapid extermination of beautiful, wild creatures. If a woman must have a fur coat, why could she not settle for a *simulated* leopard or cheetah coat? It looks practically the same. Or if she *must* have real fur, let her select a mink, fox, or chinchilla coat made from animals that are raised expressly for that purpose, instead of taking the pelts of wild animals. In our modern age, the wild animals do not stand a chance. They are not hunted with primitive bow and arrow very much. Instead, a high-powered rifle with telescopic sights is sold or given to the natives to bring in the pelts. Polar bears are hunted by plane in Arctic wastelands. Animals are shot by rifle in the Amazon, solely for their pelts. If we cut out the market for the end product, such as the fur coat, and ban the import of all such pelts, we can help to save our wildlife on a worldwide basis. Spread the word around that the spotted cats are in danger, and tell your local furrier that you will not patronize his store if he handles any coats made from spotted cats. He will get the message quickly, particularly if you belong to a women's club and visit his store with several women.

Your local zoological society can also be of great help in protecting our wildlife, if you will take an active part in it instead of just attending the meetings. You could make an annual project of purchasing small wooded lots or acreage in your area as wildlife refuges. Or you could start a minizoo, your own neighborhood zoo. Some zoological societies, such as the New York Zoological Society, aid wildlife on a worldwide basis. They finance detailed study of animals in many of the game preserves and provide funds for better game patrol to guard against poachers in the parks. They have even helped to form vast new national parks in Africa that will not only protect the fabulous wild animals of that continent but will benefit you directly, should you visit Africa as I have, to go through the wonderful game preserves there. If it were not for the diligent work of societies such as this and sincere, dedicated conservationists, both in Africa and abroad, there would not be any national parks in Africa at all. Without parks, the animals would be totally wiped out, as you can easily see if you visit Africa. Outside of the parks you will see very little game, if any.

If you do not have a zoological society in your town, start one. It only takes three people to make a club. You can have a president, vice president, and secretary or historian. I have formed several societies, including the first saltwater aquarium society in the world, which is still in existence. I also formed the Miami Zoological Society with Fred Leonard and Bob Austin, but we never expanded it for various and sundry reasons. You can easily start a zoological society or wildlife preservation society by simply getting a few friends together and outlining your plans and ideals. When you are ready for the public, send a note to your local newspaper and announce a general meeting place. You will be surprised at how many people will be interested. I hape to start a Jungle-Rainforest Preservation Society. It is sorely needed (see "Appendix A").

The main thing is to do *something* about preserving our precious wildlife before it is too late. The possibilities of helping are limitless, and the alternative is a dull world with only people fighting for each square foot of land. You might even help by encouraging the building of new cities and the tearing down of old ones. The new ones would be tall and not wide. An entire city could be built with just three buildings, each triangular in shape, for maximum viewing. The project could be set in the center of a beautiful, wild, natural area with all roads underground, in tunnels. It would have its own water and sewage plant. The waste water would be completely cleaned, and instead of dumping it into rivers, it would be used to water and irrigate the trees and grass in the surrounding area. Old cities could be torn down,

A moose in a natural setting at Alberta Game Farm. Courtesy A. F. Oeming, Alberta Game Farm, Alberta, Canada.

their concrete crumbled back to soil. Here would rise new forests, wildlife, and parks. This sounds grandiose and it would be if expected to be carried out on a national scale, but some cities could do it if a diligent society would work out the plans. Such a project could make people happier than they would be living in a small concrete box with a tiny lot of land and constant irritation with neighbors. A skyscraper dwelling with vast woods around it for miles on end, underground roads and parking, or perhaps a total ban on autos, and a single monorail for transportation to outside areas—these are big dreams, but a progressive environmental society could bring them about. The society could hire its own architect and build such a place. If we had enough cities like this we could stop destroying our remaining wild areas for "progress" and development. In today's world, progress has come to mean destruction.

It is not too late to save our environment. It would be difficult to do it on a national level, but we can do it a little at a time in our own areas. Join a club or a society and actually do something. If the club accomplishes little or nothing over a period of six months or a year, do not waste your time with it. Get out and join another group or start a new club. It will be up to the individual with purpose and drive to save our wildlife and natural areas. He must take the initiative and lead his group with new ideas and definite actions. The time to study our problems has ended. This is all that the government will ever do about it—study it, investigate it, and spend millions of dollars on research with more appropriations for additional research. We have researched enough. We do not need a team of scientists to tell us that a sewer outfall into a private lake is polluting the water. We know that already. There has been enough research on environmental destruction and pollution to cover Lake Erie with paper a foot thick. We need to act, and at least some of the above suggestions will work if carried out in a conscientious manner. We need to establish more national parks and wildlife preserves, for as our population increases, the remaining wilderness areas will be bought up by developers and turned into housing projects. I have seen many beautiful areas around Florida fall to the bulldozer in the past twenty years. It appears that some people will not be satisfied until the entire state is covered with pillbox houses and ribbons of concrete. There was even a plan to destroy a major portion of the Everglades just to build a jetport for noisy, smoke-producing planes. This would destroy the last remaining wilderness area and lead the way to the eventual destruction and pollution of all of South Florida. Our rivers and canals are thoroughly polluted now. Just imagine what they would be like with a giant airport and its accompanying traffic and businesses. Who needs it? I do not think we do. I have suggested placing it in the middle of Biscayne Bay, if it must be built in our area. At least the noise and confusion along with the fumes would not bother the people in Miami.

Rocky Mountain goat with young at Alberta Game Farm. Courtesy A. F. Oeming, Alberta Game Farm, Alberta, Canada.

If you love nature, if you like to stroll through the zoo and marvel at all the wonderful animals there, you must do at least a little something to keep our wildlife out of serious trouble. Send a contribution to your local zoo every year and support wildlife organizations. It is a fact that wildlife is in danger on a worldwide level. You can help by doing your part.

Try to initiate an open-air zoo in your home-

Black fallow deer enjoying a friendly tussle at Alberta Game Farm. Even in the far North, big-game enclosures are possible to establish, simply by choosing animals that can live in a colder climate. Courtesy A. F. Oeming, Alberta Game Farm, Alberta, Canada.

town. It is my opinion (and the opinion of others) that the present-day zoo should be phased out, if it should exist at all. As the famous Dr. Leakey put it so succinctly in his beautiful book, *Animals of East Africa,* "Keeping wild animals in zoos seems to me to be no longer justifiable, *whatever scientific benefits may come of it.* I cannot believe that anyone who has seen the superb creatures of the forest or plains, free in their ancient home, can see them behind bars without a feeling of pity."

I agree one hundred percent. I think that all present-day zoos should be closed as quickly as possible. Their grounds, which are usually public property, should be made into a park, since zoos usually protect many beautiful trees within their boundaries. Less attractive portions of the land could be sold for commercial development to raise funds for the move to the open-air zoo. Move the zoo out into the country. This is where it belongs, not in a smoky, noisy city. If your zoo is of any size at all, it should acquire a minimum of 5,000 acres for the animal compounds and exhibits. Smaller zoos should settle for no less than one thousand acres. It must be remembered that our country is growing, and that once a piece of land has been acquired and an open-air zoo constructed on it, the surrounding property will develop, so that in ten or twenty years it may be extremely difficult to purchase more land for possible expansion. Therefore, a good-sized tract of land should be acquired at the start.

Select a scenic area with wooded land, large boulders, and lakes or ponds, if available, and make use of the natural beauty without moving or destroying a single tree, if possible. If it is necessary to move out ten or twenty miles from town to get the proper site of sufficient acreage, do so rather than select a smaller area closer to town. The way our population is growing, the areas ten and twenty miles from city limits today will be part of the city before you know it. Have your zoo or zoological society purchase a large tract of land for your zoo. Not only will you be giving the animals a more pleasant place to live but you will be preserving a section of our wooded area and save it forever from the developer.

But let us phase out the present-day zoo. The steel bars and smelly concrete cages with lions and tigers pacing back and forth are prisons. We should not tolerate them as a way to exhibit animals. Visit any open-air zoo, such as Lion Country Safari in Palm Beach or Busch Gardens in Tampa, and watch the animals as they frolic in their huge, open-air paddocks. Then look at the poor, pitiful creature pacing back and forth at your local zoo and you will see a different animal altogether. The zoo director who defends these outmoded prisons, stating that his animals are contented, is living in another age. One could say that a man in prison is contented because he sits there and reads a magazine, but it would not be an accurate appraisal of the situation. Any animal would be happier in a large, natural area than it would be behind bars.

The city zoo must go—if not all of it, at least the biggest part of it. It might be more practical to keep a portion of the zoo at its established location for research or temporary confinement of animals, particularly newly imported animals that must be treated and observed before they can be released into their new home. Also, perhaps parts of the zoo that have large, outdoor paddocks could be retained at the old area to keep a little of the zoo in town for the convenience and enjoyment of those who could not make the trip out into the country. Some of the buildings could remain as administration offices, or hospital treatment areas for sick or injured species. But the main part of the zoo with the big animals—the elephants, rhino, hippos, giraffes, chimps, and monkeys—should be moved to the country where the animals can have wide open spaces to explore and enjoy.

There is hardly an animal lover alive who would not rather watch animals running free in a large spacious hammock than see them confined in steel and concrete. With a little ingenuity and imagination, every species of animal, including even full-grown elephants, could be safely housed in a large, natural setting. I recently saw a full-grown tiger living in an open hammock at Lion Country Safari. It was sitting on a knoll with a half dozen lions. I thought I was seeing things until a local guide, Jim Smith, drove up in a zebra-striped truck and explained that the tiger was indeed real. He said that it had been injured and was near death and had been given to the exhibit, which nursed it back to health. The outdoor living no doubt had a beneficial effect on it, for it is a superb animal. I am sure that the tiger was much happier there than it would have been behind bars in a circus or a zoo.

Winter snow and cold weather, of course, will have to be considered in northern climates, but most tropical animals can be acclimated to the cold to some degree, as has been proven in the zoo at Switzerland as well as at the Catskill Game Farm in upper New York and the Alberta Game Farm in Canada. Some zoos let their animals frolic in the snow and the animals love it. Of course heated quarters would be needed for the outdoor zoo. It might be feasible to run heated water through pipes underground and keep at least a small area in each compound frost free and green, even during the coldest weather.

Those cities with power plants using water as a coolant might make use of this heated water to warm the zoo and surrounding lands. The so-called thermal pollution thus could be put to good use. It could become a real asset to purchase land near the power plant and make use of their hot-water discharge. The heated water would be free and could heat up small ponds or lakes for year-round use by the animals. Even the grass and surrounding trees could be sprinkled with the warm water in winter, keeping them frost free and green and providing both the animals and the people who visit the zoo with a "garden of springtime" all through the cold winter months. It is certainly worth investigation and consideration. The city dweller could have eternal spring at his doorstep, quite a refreshing change from a cold, wintry day.

I suspect that this book will have many critics, particularly from a few of the old-time zoos. Some zoo directors will categorically oppose the keeping of any exotic animal by Mr. Average Citizen. Yet it is a fact that they, themselves, kept all kinds of exotic animals in their earlier years, which led to their career as zoo director. They have all kinds of animals at their disposal and can take baby animals home with them whenever they please, but no one else is supposed to do it. This is not quite right. They may also state that the open-air zoo is not practical, that there is too much danger of the animal escaping or harming someone. This is negative thinking, for an animal is less likely to want to escape if given a huge outdoor hammock with a large, rock house and trees in which to live. The outdoor hammock can be constructed so that the animals cannot escape, and they are less likely to get out than when confined in a small cage where any weakened area is quickly enlarged by an animal that has nothing else to do *but* escape. As for people being harmed by the animals, again this is less likely in a large area, because the animals are not in close contact with the people as they would be when the animals are in a cage. While there are a fair amount of injuries to people in the present-day zoos, such occurrences are almost nonexistent in the open-air zoos. Even in the wild-game preserves of Africa it is extremely rare for a person to be harmed in any way by the animals, unless he is downright foolish. As long as you stay in your car the animals will not bother you at all. The same would apply to the modern open-air zoo. People could drive through the area at a slow speed of ten to fifteen miles per hour and stop if they pull to the right so that the car behind them can pass, if necessary. None of the animals will bother them if they obey the rules of the exhibit, which will vary according to the area and species exhibited.

The open-air zoo is the zoo of today. It's the only way to go. I am sure animals lovers, animal protection groups, and all those who love and enjoy the wild creatures that share our planet with us will agree. It is not only more enjoyable for the animals whose lives we hold in trust, it also protects forever a section of our vanishing wilderness. We all need that. But we need more than zoos to protect our threatened wildlife. We need more of the huge, natural animal preserves, like Kruger Park, Serengeti, and the other large game parks of Africa. But we need some in the United States and South America, especially where large tracts of virgin land are still relatively untouched. Whether he realizes it or not, man will survive best if he lives close to nature. All his marvelous inventions and creations are nothing when compared to the beauty of a single tree. No artist's painting, no architect's building, including the great pyramids, not even a poem is more lovely than a tree, for the tree not only gives beauty, it refreshes the air. For this reason alone, it is important to man's very survival, regardless of how selfish he may be. If man were to flood the Amazon and kill every tree in this vast wilderness (a project frequently proposed), he would not triumph in the end, for he would find that his very existence has been destroyed. The long chain of ecology is linked to the forest and the wildlife. Destroy them and man destroys himself.

Many scientists and naturalists now feel this way more than ever before. At a recent architect's meeting, it was proposed that legislation be adopted to protect every single tree in the land and that a tree could not be cut down just to erect a building. If there were no way to erect the building without harming the tree, then the tree would have to be dug up and replanted. This is a marvelous idea and one that could well be adopted within the decade if our country moves toward peace instead of war.

Only by protecting the forest, the veldt, the jungle, and the tundra with national park status for vast areas, can we insure the survival of our remaining wildlife and the reestablishment of our endangered species to a safe level. The wilderness parks are *not* a waste of real estate, as some might feel. It is a fact that millions of people visit the national parks every year in our country alone. We all owe heartfelt thanks to the conservationists who had enough vision to promote and establish our national parks, for if it were not for them, the last of the giant redwoods of California would now have passed through the sawmill and the Grand Canyon would be dotted with hotels and high-rise apartments. But we need to establish many more national parks, for our present parks are becoming extremely crowded. We need them in every state of the union and in every country of the world.

A few of the new parks should be established to protect exotic animals that are threatened in their natural lands. The Javan, Indian, and Sumatran rhino, the gibbon, and the orangutan should all be brought to our country and released in special wildlife preserves in warm area. The Siberian tiger could be established on an island off of Alaska to insure its survival. While I do not propose to stock the entire country with elephants and rhino, I do feel that we already have millions of acres of national park devoted to preserving our own native species. We need at least a few special areas to preserve severely threatened animals that may be totally killed off in their native land, should the present government there change hands to a less conservation-minded government. A good example of this is in the Congo, where thousands of elephants were slaughtered in the national parks when the country was given its freedom. Most of the animals were simply left to rot after their ivory tusks were removed. Even the rare rhino were killed.

We need to bring a nucleus of endangered animals to our own country so that they can be established in special national parks where they could be protected and kept under constant observation. A minor tourist charge to the park would help pay for the transporting of the animals and maintenance of the area. An example of animals needing such protection is the Indian rhino. Poachers are still killing them in their native preserve, and an uprising or war in the area could easily result in the total slaughter of all of the remaining animals. I would propose that sixty of the animals be brought to America by our government and established in three groups of twenty each, in a special rhino preserve: one in Central Florida, one in Southern Texas, and the third in Southern California. The animals could be released in a five-thousand-acre sanctuary in their prospective areas, and in all probability they would reproduce to proportionately larger herds in ten or fifteen years. The day may well come that the offspring from these animals could be exported back to their native India to restock areas where they once flourished. The same could be done with other animals. We could establish interesting wildlife parks in our country, and at the same time it would not only protect the trees and wild areas but we would safeguard against the possible extinction of many animals. It is not enough to protect animals in their native land, for it is too difficult to patrol the areas. In the swamps and forests, anything can happen, and poaching of animals is often carried on with impunity. No one really knows how many animals are left of certain species, as it is difficult to count animals in a jungle or a swamp with fifteen-feet-tall grass. There may be hundreds of animals there, or there could be just a dozen or two, which could be slaughtered in an afternoon.

We must protect our dwindling wildlife through more and larger zoos, wildlife parks, and preserves, and we must educate our children, so that even cutting down a single tree would be cause for serious evaluation and consternation as to its ultimate effect upon our own ecology. Further, we must introduce herds of exotic animals from other lands, including even the elephant and rhino to at least some of our national parks. In a recent article in *Animals Magazine,* London, England, noted conservationists were asked to name their favorite national park. Sadly, no United States parks were chosen. Instead, nearly all chose the African parks because of their abundant wildlife. In our own national parks (I have been to most of them) you are fortunate to see even a squirrel, but in Africa's parks you can see animals in every direction. We should have the same here and it can be done.

Appendix A

A Tentative Plan to Protect the Jungles
The Jungle-Rainforest Preservation Society

By Robert P. L. Straughan

An Urgent Message!
S.O.S. The Jungles Are Going! The Rainforests Are Dying! The African Veldt Is under the Plow! S.O.S. Help!

Did you know that in South Africa, you are able to travel from Johannesburg, all the way to Kruger Park, and see no wild game whatsoever? The once beautiful veldt is now cornfields, orange groves, and sugarcane. All game has been killed off right to the very boundaries of the park.

Did you know that on the way to the famous Treetops in Kenya, you will find barbed wire fences, tractors plowing land and cornfields, right up to the very gates of the park?

Did you know that the road to the famous Ngorongoro Crater, a fabulous stronghold of wild animals, is dotted with farms and houses almost to the borders of the park itself? In fact everywhere one travels in Africa, there is evidence that man is taming the wilderness, mowing down all the trees and planting the once magnificent veldt in neat rows of crops. Even the Ngorongoro Crater itself was considered for a garden by agriculturists.

The giant Amazon is now under attack. Huge areas of wondrous rainforest are being slashed and burned to make way for crops. There is a road being built across the center of the Amazon basin itself to "break its back" and civilize this most marvelous creation on earth, a vast, mysterious area that holds a large portion of the world's mammals, reptiles, birds, and trees.

You may not even care and you may have no wish to ever visit any of these areas, but perhaps your children or grandchildren might. But how can they if the areas no longer exist? They can't! It is up to you to help preserve the wild areas of the world by *buying* it. Money is the international language and it is money that will save or destroy the beautiful wilderness areas on this earth. Let those of us who love nature and the outdoors preserve it forever by simply buying it.

Individually, it would not cost you much, a mere ten dollars a year or more if you can afford it, but if we can get fifty thousand or even a million people to help, we can buy a vast rainforest, a beautiful veldt area, or a mysterious jungle. All you need to do is join the Jungle-Rainforest Preservation Society to help accomplish this marvelous deed. You can further help by getting five or more new members. If you do this, your next year's dues will be automatically paid. You will be doing something not only for yourself, but for all mankind, for once the jungles are destroyed, they are gone forever. Once a

Gazelle at Serengeti Park. You will see more animals at a glance in any animal preserve in Africa than you would in all of the national parks in the United States combined. We could change this by bringing a half million or a million animals from Africa to America and releasing them in large, fenced areas of land. Many animals in Africa are doomed, even in the preserve, and this would be one way of saving them.

species of animal dies out due to lack of its wilderness home, it can never be replaced. We will have lost a part of the living earth.

Adopt the motto, *"Buy the jungles, save them forever."*

Help us buy the jungles and rainforests. You will get many benefits, including the right to some day visit a jungle area you helped to purchase. You will get a semi-annual or annual report of our progress and eventually an illustrated magazine showing you actual pictures of the wonderful areas you helped to preserve. You will also get a full account of every cent of the funds spent for advertising, officials' or rangers' salaries (to be kept painfully low), etc. Your money will not be squandered on plush, high-paying jobs or million dollar magazines. As much as possible will go to the actual purchase of land and the yearly taxes and upkeep of same.

Wouldn't you like to be part owner of the mighty Amazon jungles, an Ecuador rainforest, a jungle island in Indonesia, a big-game preserve on the African veldt, or a forest in the Congo? You can! Read on!

Beautiful gazelle in one of Africa's game parks. The author suggests that giant new parks be established, specifically for African animals. It would add great interest to our national park system and would help preserve the dwindling wildlife of the world.

The Jungle-Rainforest Preservation Society

"Dedicated to the permanent preservation and protection of pristine jungles, rainforests and wild areas, particularly in the tropics. Operating on a worldwide basis to save our diminishing wildlife from the encroachment of civilization."

It is not enough to protect the animals themselves. Their habitat must be secured and protected not only within its boundaries, but in all areas that lead to the preserves. It makes little sense to save a wild area only to have it ravaged by an oil spill or destroyed by noxious fumes from a nearby pulp mill or chemical plant.

The Jungle-Rainforest Preservation Society has been founded by Robert P. L. Straughan, author and naturalist, who is concerned with the plight of wild animals and wild places. This is an active society with a strong purpose, which is to actually buy or lease for indefinite periods, vast tracts of untouched and even unexplored jungle to protect these areas from the slash and burn farmers, the engineers with their damn dams that flood and kill everything in their path, and the land developer with his bulldozers and dredges. The land will be purchased before it is destroyed and saved forever for wildlife specialists, naturalists, and nature lovers to enjoy.

It will never be developed while in our possession, and roads will be strictly taboo as will all types of construction with the exception of an occasional forest lodge and tenting encampment to be kept on a small scale for visitors to the preserves. These will eventually be opened to members at a modest fee and to nonmembers at a higher fee, the profits and proceeds to be returned to the society fund. This will enable members to actually visit the wild, unspoiled jungles and rainforests that we will guard and protect for all mankind before they are utterly destroyed. Access to jungle camps will be made by river or forest trail by Land rover, in such a way that no scars are made upon the primeval wilderness.

Unlike other conservation societies that may spend millions of dollars on a fancy magazine and high salaries for their staff, JRPS will hire a limited personnel at a modest salary and may eventually publish a low-cost publication to keep members informed of the work and progress of the society.

We will publish an annual or semi-annual report accounting for every cent of funds collected so that members can see exactly how their money is being used. To our knowledge no other nature group has done this, and members wonder if just a trickle of funds actually reaches the final aims of the society, which is often the case.

JRPS is founded with *integrity*. Its purpose is urgent and sincere and should be of great interest to all those who love nature, wild animals, birds, and, most of all, the untouched wilderness. By joining, you will actually own a bit of the mysterious jungle and be preserving it forever. Isn't that exciting? You will be placed on our mailing list to be sent items of interest as they occur so that you may keep abreast of what is happening. As the society grows, we hope to put out a *modest* magazine, mostly black and white, to show you actual photos of the jungles and rainforests that your money has purchased, as well as a *full account* of all money spent by the society. Funds sent to JRPS will be hoarded carefully. The money will *not* be spent on lavish publications or high salaries. Salaries, legal fees, travel, and purchases will be done where we feel we can get the most for the least spent but still get good results. There will be some voluntary work and this will always be appreciated, but we feel that for a long-range project it is best to pay our way so that we can attract and keep persons of high caliber and sincerity. A man cannot work for nothing.

A Few Questions about the Society

QUESTION: Where will jungle lands be purchased?

ANSWER: In Africa, Asia, South and Central America, and even in our own United States: anywhere in the world, but mostly in the tropics.

QUESTION: How large will these areas be?

ANSWER: This depends how much money is brought in annually, how fast our society grows, and the availability and security of the land. It may encompass five hundred to five thousand or even five million acres. We will seek to obtain lands that are clearly defined by rivers, mountains, and seashore and that are uninhabited so that there can be no future claims on them.

QUESTION: Who will designate the lands to be purchased?

ANSWER: We will have a board of directors who will carefully evaluate properties to be purchased or leased in perpetuity. We will also eagerly look forward to hearing from all members, particularly those who travel, for their ideas and suggestions with priority on the immense rainforests of the Amazon, the Congo, and Central America.

QUESTION: Will hunting and fishing be allowed in the preserved areas?

ANSWER: Yes. Unlike many nature preserves where hunting is prohibited, we may permit *limited* and carefully regulated hunting in the preserves under special license and only for species that are in excess or damaging to the environment. No hunting would be permitted near forest lodges or tenting areas, except for food for the table, and this would be taken in the farthest extremes of the immediate area. Proceeds from hunting licenses will be added to the general fund. Fishing will be permitted (with license) in most of the preserve areas except around tenting areas where it would interfere with the natural viewing of wildlife.

QUESTION: Will live trapping and collecting of animals and specimens be permitted in the preserves? Will logging be permitted?

ANSWER: Yes, depending upon the size and location of the preserves, *limited and regulated* trapping of certain species of animals for zoos or other purposes will be permitted, provided that the species collected are numerous in the area and that their removal would not noticeably affect the general population of animals in the preserve. It has been scientifically proven that any wild area can only support a given number of animals and sometimes control or removal of the animals is more humane than allowing the animals to increase to plague proportions and thereby destroy both themselves and the natural habitat. All profits of course would be returned to the society. Logging is prohibited!

QUESTION: How will the money be handled and spent, particularly at the beginning?

ANSWER: All funds collected will be placed in a savings and loan account to draw interest, with a small portion for a checking account. After a year or two, if insufficient funds are collected to carry out the aims of the society, all money will be returned to the contributors, or sent in a lump sum to a selected conservation group after expenses for printing, stationery, legal fees, advertising, etc. have been paid. But we hope this will not be necessary and we will keep you informed of our progress, hoping you will eagerly solicit members for us. Remember, it is your society. You will be buying and preserving a bit of jungle for yourself and future generations to enjoy.

QUESTION: What about the Rainforest Lodge. Will I be able to visit it?

ANSWER: Yes, but at your own expense. As a member you will be given a discount at the lodge as well as a preference for reservations. You will also be given a discount on hunting and fishing licenses in the preserves. But do not

Antelope at Serengeti Park. A habitat such as this would be simple to duplicate in the southern areas of the United States by running a double, ten-foot-high chain-link fence around the area and turning the animals loose inside. Periodic culling of excess stock could produce animals for zoos or for release upon public lands.

expect us to build a rainforest lodge right away. This would cost a lot of money, not only to build, but to maintain. It is a future goal of the society so that members may enjoy the fruits of their efforts. Jungle lodges would be built on the order of the famous Treetops, animal viewing lodge in Africa. They would be built high up in the trees of a rainforest, far from civilization, with feeding stations to encourage the wildlife in the area to show itself.

QUESTION: How can I become a member of the Jungle-Rainforest Preservation Society?

ANSWER: By sending in your contribution and filling out the enclosed form, you will become a member and you will be actively preserving a real jungle or rainforest.

If We Are to Have Wild Places Left in the World Today . . . We Must Succeed. With Your Help, We Can!

This is not a scheme to take your money. We don't want it or need it for ourselves. This is a sincere and genuine plan to buy up the jungles before they are destroyed forever. We may arrange to name preserves after generous donors if this proves feasible and we hope to become a perpetual society, worldwide in scope. Send in your donation and buy a jungle.

NOTE: The above is only a tentative plan under consideration by the author. It shows one way in which our jungles and wild areas could be preserved within our lifetime for future generations. It is included in this book only as a guide and suggestion of what is needed. Perhaps someday, the author or others will form such a group. It would probably work best as a nonprofit wildlife foundation, but it could help protect wildlife and wild places on a worldwide basis.

Appendix B

Conservation of Endangered Species Will Not Be Accomplished With More Legislation*

Opinion by Alton Freeman

Conservation is a broad term. Conservation of nature is in itself such an ambiguous term that the average person isn't quite sure just how it should be applied. When you get down to conservation as applied to wildlife, here is a specific that, in general, means to keep alive and reproduce the wild creatures of the world. A large segment of our population, if they have given it any thought at all, sort of passively agree that this type of conservation is commendable.

There is, however, a much larger group, including many hunters, the furriers, hat makers, lumber and other natural resource exploiters, who care little or not at all whether wild animals continue to exist . . . their interest, in part understandable, is for their own immediate pleasure or business purpose. And, finally, there is the majority of the world's population who have no thoughts at all on the subject.

Citizen's Knowledge

The average person living in a large city is not a hunter, has no exotic pet animal, has no contact with mining, oil exploration, timber production, or any other natural resource industry . . . his only knowledge of wildlife comes from an occasional visit to a zoo where he admires the gorgeous colors of an exotic bird, the comical antics of a monkey, size of an elephant, or ferociousness of the big cats. Such a person is not concerned with the fact that the Bongo or Okapi is in danger of extinction . . . after all, unless there is a sign on the

---

* Reprinted from *Pet Shop Management,* February, 1969, 1 East First Street, Duluth, Minnesota 55802.

pen, a Bongo or Okapi is a "deer" to them, and there are all sorts of deer aren't there?

Animal lovers (of which I am one), scientists, educators and zoologists as a rule, do want to conserve our rapidly diminishing wildlife. We believe that a live bird or animal contributes not only to our knowledge of existence, but to the pleasure of seeing something alive so different from a human, as well as an opportunity and challenge to express our affection to a nonhuman. And if this affection is returned, is there any greater reward to such a person? Is this not proven by the immense dog, cat and bird pet population kept in every nation on earth?

We agree, therefore, that conservation of wildlife should be a necessary part of humanity. We do not agree, however, on just how it should be carried out. It is my belief that far greater education of the passive masses is necessary. One way to effect such education is, of course, to encourage more people to know more about such creatures. Surely the best way to know is to see, if not "to have and hold." Our expanding civilization makes it impossible for most people to "have and hold" a wild bird or animal, thus a zoo is the only answer, and to a lesser extent, the flash photos of TV which never concentrate on one interesting exhibit long enough to really know it, or which fake actions to such an extent as to give an entirely erroneous impression.

The American Association of Zoological Parks and Aquariums (AAZPA) has passed a resolution that certain rare species will not be purchased by them in order to conserve the world's rapidly diminishing wildlife. At present the list is not too great, but does include such interesting animals as Orangutans, the Galapagos Tortoise, etc.

Pressure has been put on our USD Interior by various groups which caused the Lacey Act to be passed and enforced to almost disastrous proportions. This Act prohibits the importation into the U.S. of any bird or animal whose exportation is prohibited by the country of origin (when the jungles of one country with a prohibition adjoins another who permits, or when an exotic animal is bred in captivity by a distant country, this is somewhat of a problem to prove).

U.S. Programs

Our Department of Agriculture has a long list of restrictions and prohibitions of mammals, mostly wild ruminants. Health, Education & Welfare Department restricts some species. A special act of Congress bans the Golden Eagle. Many States prohibit certain species, and most major cities have their own laws regarding wild animals. The Plant Board, USPH, USDA, USDI, Customs, etc. all inspect every single importation and apply their regulations to each. To these prohibitions will soon be added 500 additional animals claimed to be in danger of extinction.

The country of Brazil last year passed laws prohibiting the exportation of all wildlife, thus no longer can even a parrot be purchased in a Brazilian Pet Shop. Colombia, Venezuela, Peru and Ecuador will meet this month to discuss the prohibition of all of their wildlife. Such countries as South Africa, India, Indonesia and even tiny Nepal rigidly control exportation of their immense numbers of interesting wild birds and animals. The handwriting on the wall is that zoos all over the world, but particularly American societies, will have their stock severely reduced, if not completely eliminated in the foreseeable future.

If this new long list of prohibitions is passed by Congress, it is probably that pressure from zoo and laboratory groups will try to have their stock excepted by means of permits and/or other controls. While this would appear to be a reasonable solution for these two groups, the end result will simply mean even more voluminous red tape than already exists as well as a nearly impossible increase in costs to such an extent that only a wealthy zoo or laboratory can afford to import what today are the more common-type inexpensive mammals.

Livestock Sources

Nearly all zoos and labs purchase their livestock through an Animal Dealer as the problem for them to import just one mate or one pair of animals is too great and too expensive. At present these Dealers import either by direct order from the zoo, or more often, by speculation of a sale to them or to a laboratory or pet shop. An Animal Dealer, like any other businessman, must make a profit, and to do so he takes the considerable calculated risk of investing in a product for which his experience indicates there will be a demand. By importing a group of several hundred small Squirrel Monkeys, for example, the costs of catching, buying, shipping and holding are reduced and a variety is available so that there is a stock on hand for the immediate filling of a laboratory order requiring two, three or twenty animals for a continued experimental need.

Such a Dealer may also have a mate on hand for replacement to a zoo who has lost one of a pair, or an immediate supply for a new Children's Zoo or an empty pen. But, if new laws require a zoo or laboratory to obtain a permit for each single replacement or small order and a Dealer is not permitted to keep a stock on hand, there simply will be a strangulation of supply: The exporter is not going to go through the rigmarole of red tape to export just one inexpensive specimen, the catchers are not going to bother holding one inexpensive specimen for some indefinite time, the Dealer is not going to bother with correspondence and paper work for each individual small order unless compensated for it by sales price, and the transportation from abroad is considerably more for a small shipment than one of volume (for example, it costs no more to ship five or six small monkeys, 20 to 30 small birds, than it does for a single one).

The result of all of this would mean the immediate curtailment of laboratory specimens and exotic small zoo stock. Added to this curtailment will be the almost impossible addition or replacement of stock to private breed-

Tigers are becoming extinct in many countries of Asia. There are said to be only a few thousand left in all of India. Zoos have a good record for breeding them, such as this fine cub born at Crandon Park Zoo. Perhaps someday we will export *animals to Asia to reestablish them in their native land. The author feels that rare species like the giant Siberian tiger could be released on uninhabited islands off of the Pacific Northwest, where they could reestablish themselves.*

ers in this country (which hopefully may not be included in the pet ban). These individuals are currently doing a most successful job of producing many exotics of waterfowl, antelope, etc. . . . even more so than most zoos, even under existing nearly impossible regulations.

In my opinion the Wildlife Conservation people, including the AAZPA, are incorrect in the thinking that by leaving a rare bird or animal to its natural environment in the wild will cause it to become more plentiful. In the first place, if a specimen is "rare" there must be some reason for it caused by the natural environment. . . . Either it has become depleted by other animal enemies, the habitat has become barren for one reason or another causing a scarcity of food, a disease has killed it off, hunters have decreased it, or, as in many cases the simple expanding encroachment of civilization has depleted it. Thousands of Kudu and other antelope have died in South West Africa because of drought; thousands of flamingoes die each year from salt-encrusted wings; fire in many wild areas does its bit; poaching everywhere is a problem. For most of these causes, we in this country can do little or nothing, so leaving the bird or animal in this environment is not going to increase the rare species, but possibly help destroy it.

Some Successes

True, we in America saved the American Bison by leaving it to itself . . . (but also by prohibiting American hunters to shoot it) ; we helped save the Bird of Paradise by prohibiting the use of the feathers in the millinery trade; we are currently saving the Whooping Crane somewhat by protection (but primarily by breeding in an American zoo and restricted sanctuary). BUT, that is this country! Many other countries are not so successful enforcing protection laws. The vast jungles of Brazil cannot be policed. The commendable East and South African Departments of Conservation are not stopping poachers. The overly-protective laws of Australia actually cause an over-population of certain species such as Kangaroos and Cockatoos that menace farming. Birds-of-Paradise feathers can still be bought in Singapore shops just as Rhinoceros horn (used as an aphrodisiac by the Chinese) is available for buyers. The pelts of Snow Leopards, Lesser Pandas, Polar Bear, Clouded Leopards and other rare animals can be bought in India, Kashmir, Nepal. . . . Even New York, Paris and London sell exotic animal coats. The trophy rooms of American millionaires proudly exhibit the heads of "rare" hoofed stock . . . the more "rare" the more boasting.

All of these factors make the job of conserving wildlife in its natural state almost impossible . . . thus, would not the most sensible manner of conserving be to allow the importation of any exotic animal or bird, not only for the benefits it would give by exhibition, but also for the purpose of reproduction? We already have many laws to protect the inhuman treatment of animals (if this is part of the argument for prohibition), and in general there is little inhumanity as such in this country . . . nothing to what it is in most exotic lands of origin.

A recent Life magazine story stated that in one year the U.S. imported 28 million live creatures, 75,000 of them mammals. A representative of the Department of the Interior also stated that because of this Life article which was most critical of the animal importations, they had to put on dozens more personnel to answer the volume of correspondence from the public about it, and this was one of the contributing factors in the Department's continued plans for additional exclusion of live animals. Such figures are, at first, impressive, but when one analyzes them what do we get? I don't know exactly how these figures are computed, but assume that a good part of the 28 million live creatures were fish (and I don't hear anyone banning them) ; and of the 75,000 mammals the majority were laboratory monkeys. It is claimed that such importations are depleting the jungles.

Conclusions

The rule of economics will, in general, balance out any abuse of importation of rarities by the simple fact that the more rare an animal the more expensive. Thus, few people can afford the cost of such simply as a pet, and the few who can are doing so with the natural intent of properly caring for it and/or attempting to reproduce it in captivity . . . it would be foolhardy to do otherwise. If a family buys a $20 monkey or bird as a family pet, it stands to reason they do so because they have an interest in owning it. It also is reasonable to assume that because the price is low there is an abundance of that particular species; therefore, there is reason to believe that the bird or animal is serving a greater use to humanity by being a pet than it would fighting for its existence in the jungle. Should the supply of such pets become scarce, then the price would naturally increase to make them unavailable as such, which obviates the need of any additional laws of importation prohibition, does it not?

I say we already have enough wildlife conservation laws in this country, and what we need is a changed attitude of the meaning of the word and a more effective promotion of control towards those who really destroy wildlife not those of us who promote its continuation. Let American zoos import every living natural bird or animal available which they can properly exhibit, control and reproduce! Permit private American animal breeders to import and reproduce any specimen which has a chance of continuing its existence. Allow laboratories to have a sufficient quantity of experimental animals available at lowest costs for the most necessary investigations into human, as well as animal health. And lastly, do not stop the supply of pets to American homes which by the very nature of it can teach affection, care and understanding to our children.

9
The Old Zoo and the New

Should we abolish all zoos? Occasionally this question is asked by conservationists who feel that it is cruel to keep animals in a cage and that the animals would be better off if left in the jungles. They have a good point. Animals would probably be better off if left in the wilds, but then perhaps the whole world would be better off if there were no people in it. We have certainly made a mess of it. All kinds of things are not quite ethical so far as nature is concerned. Does man have the right to chop down mountains so that he can get at the minerals inside? Does man have the right to draw all of the oil from the earth, polluting our rivers and oceans just for the gain of a few? There are pros and cons for everything, zoos included. Any zoo director will tell you that zoos are great and that they are making a real contribution to science and nature. What else could they say? Perhaps a more logical question would be: Is there no better way to exhibit animals than behind bars? The answer here is *yes.*

We need zoos, for people should be able to view all of the strange animals of the world. It is a refreshing change from just looking at people. Zoos are very important, and the wealth of information that they have accumulated throughout their years of existence have given us a much better understanding of wildlife. Modern zoos are now striving to breed animals within their own stock rather than require new animals from the wild, thereby reducing the demand for trapping wild animals. In fact some zoos are increasing their breeding stock so that surplus animals can be supplied to other zoos. Endangered animals are often entrusted to zoos, where expert care and constant surveillance will ensure not only survival of the species but an increase in the population of the species as well. This is of course a splendid effort, and zoos both large and small, public and private, can benefit the animal popula-

Does man have the right to imprison an animal just so that he can look at it? There are ways to exhibit animals other than steel bars and concrete that are used to confine this unfortunate gorilla. Gorillas and chimps can be exhibited on islands surrounded by a water moat. Fortunately, many zoos are doing away with the steel "prisons" that belong in the Dark Ages.

The prisoner. What has he done to deserve a life sentence behind bars? It is scenes such as this one that make the new outdoor zoo a necessity. This baboon deserves a huge outdoor paddock upon which to roam and play.

tion by the continued propagation of all animals held within their confines.

Surplus stock from some zoos and animal farms is sometimes released on public lands, particularly hoofed stock from Africa and India, which are so closely regulated by our government. A few state conservation agencies have already embarked upon projects to introduce African antelope into their wilder areas as a game species. One animal, the Barbary sheep, which was introduced to New Mexico, did so well that limited hunting of them is allowed to keep the animals in check. New Mexico has also been experimenting with kudu, gemsbok, and impala. The animals were brought in with cooperation from zoos and kept in large enclosures, where they would be allowed to reproduce on land similar to that in which the species was to be introduced. Only the young of imported hoofed stock are ever allowed to roam free in our country, as a safeguard against disease.

A number of states are now working in this area of wildlife conservation, the idea behind it being the fact that, in addition to providing a fascinating new animal to our native wildlife list, there is the safety margin in preserving the animal from extinction, should it ever be killed off in its native land. In Texas, the Nilghai antelope from India is now established in vast herds said to

The magnificent tiger, a prisoner of concrete and steel, Animals lovers cannot help but feel pity for caged animals that must pace back and forth with monotonous regularity for their entire life. It is hoped that our new zoos will give the animals room to run.

The caged ones, such as this monkey, deserve a better place to live.

Kangaroos at the zoo. They would be a lot happier hopping about a giant, grass paddock of ten acres or more.

number up to a thousand or more on private ranch land. Some of the wealthy ranch owners have set aside huge tracts of land for exotic animals. They of course must work closely with the zoos to bring in the animals. Some of the larger ranches even have elephants and giraffes roaming free on their vast acreage. As you drive through the area you almost feel as though you were in Africa. I am always pleased to hear of such projects. It is something I would like to do myself some day. I write to various conservation departments from time to time urging them to consider the introduction of at least a few African animals to our lands. I usually get the English sparrow and Australian rabbit story retold with monotonous frequency, but occasionally I get an encouraging letter saying that at least one state is considering releasing something exotic other than a pheasant. I point out to these departments that such animals as the white rhino, the ostrich, the giraffe, and the hippo could live quite well in the southernmost areas of our country.

Surely this bear cannot be even remotely happy in its steel prison.

The zoos work very closely with conservation departments as well as with private ranchers who wish to keep exotic animals on their land, particularly if the rancher agrees to buy animals for the zoo and will accept the offspring of the animals for release on his land. Government regulations on importing hoofed stock are extremely strict. The average citizen cannot import them at all, since they can only be shipped to an accredited zoo where they can be held under quarantine for the rest of their lives. Only the offspring of these animals can be sold to individuals to be exhibited or released upon their land. This is to guard against hoof-and-mouth disease, anthrax, and other diseases, which can spread rapidly and prove disastrous to domestic livestock. These rules do not apply to all animals. In fact hippos, rhino, elephants, lions, and many of the large animals are exempt. It applies mainly to cloven-hoofed animals, such as antelope and swine, but the rules could be broadened eventually to include many more species.

Each year, we see more and more regulations concerning wildlife. In 1972 there was a fantastic increase in both federal and state laws on wildlife. The vogue seems to be that no one except zoos should be allowed to import and display animals, and I am opposed to this. I think zoos are fine places as a whole. Most of them take excellent care of their charges, but I do feel that a number of them could do better than confining their animals to steel and concrete. I feel that many cages in our zoos are woefully inadequate and should be torn down. I of course advocate the

Unfortunately, the old zoo concept with its steel bars and concrete was used in constructing Miami's zoo. This bear cage is an example of the old zoo concept. A new outdoor zoo is being planned, and it is hoped that cages like this one will be left behind when the zoo moves to its new location.

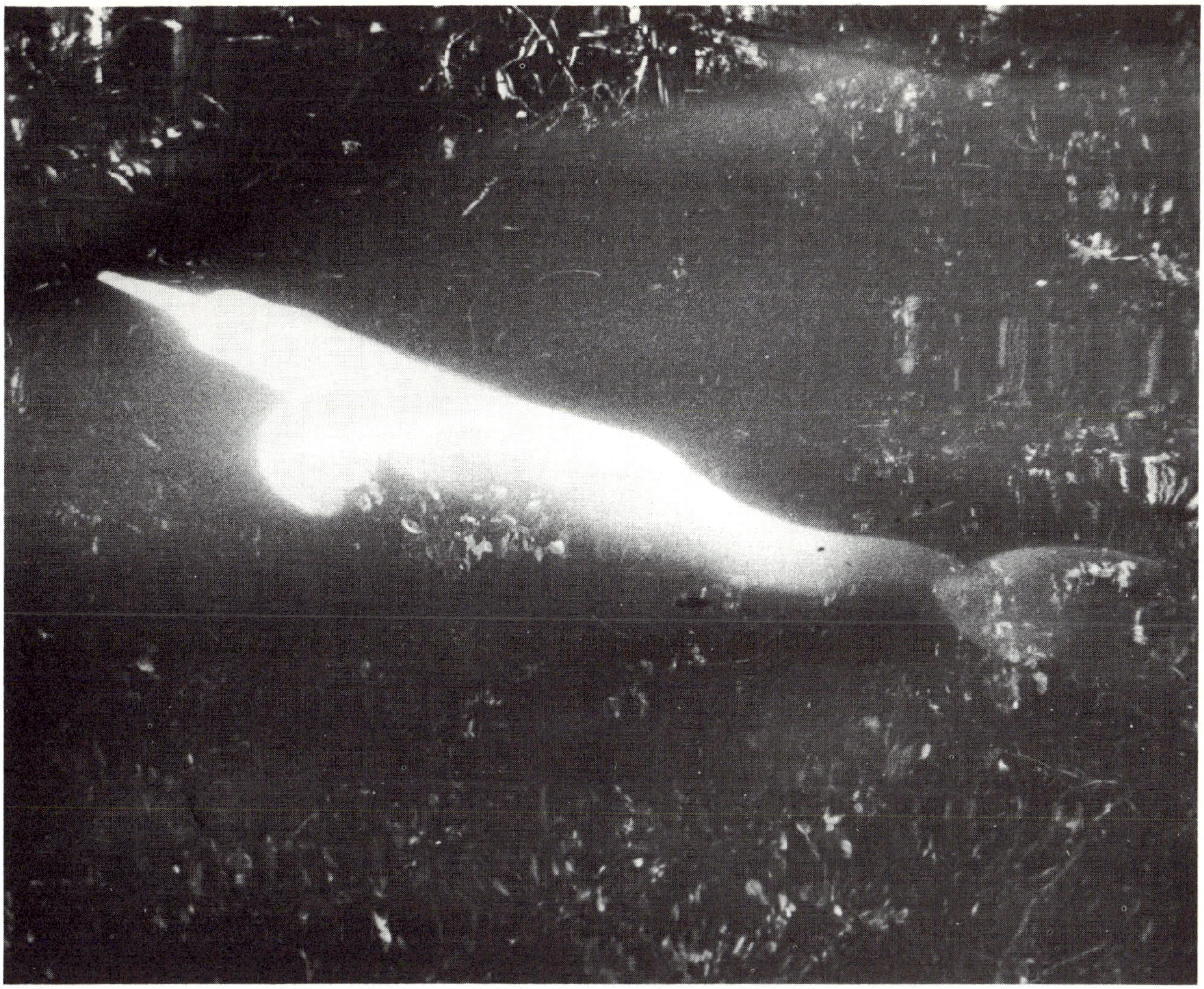

Like a ghost gliding through a dark lagoon, this spectacular white, freshwater porpoise was one of the most interesting animals at the Crandon Park Zoo. Unfortunately, it died during a hurricane when several feet of water swept across the zoo grounds from the nearby ocean. This is the main reason for the zoo's anticipated inland move to safer grounds.

The white porpoise had a little lagoon all to itself, making an ideal place to exhibit it. This valuable animal would do well in a giant outdoor zoo in a natural pond with many shade trees. It would probably reproduce under such conditions.

open-air zoo, in which all animals are kept in as near a natural condition as possible. There is a trend among zoos to keep animals this way, but it is progressing too slowly. Private zoos seem to be making more rapid progress in this field, perhaps because it is less complicated politically.

The so-called modern zoo with its concrete cages and steel bars is as outmoded as the horse and buggy. It has served its purpose well, for it has brought a little of the jungle to the city. Today's animal lovers cannot help but feel pity for the "prisoners" in their jail cells. What have these animals done to deserve a lifetime of incarceration? Is it any more moral to place an animal in a steel jail for the rest of its life when it has done no wrong than it would be to put a human behind bars for no reason at all? We know it is wrong.

A rare white tiger, on display at Crandon Park Zoo, is one of the most magnificent of all animals. This unusual specimen was obtained for the zoo by Ralph Scott. A mate for the animal was secured, but before it was delivered to the zoo, the tiger was found dead in its cage. It died of a viral infection. The zoo had hoped to breed the beautiful animals.

We are living in changing times of space technology. Man has already gone to the moon, and the planets are awaiting a visit. Color television has replaced black and white. Instant coffee has replaced the home brew. Everything has changed except the zoos. They are still in the last century. In their day, they were acceptable as the only way to keep and display animals, but we are in the space age now. Many visitors to the zoo look with disdain at zoo officials, upon seeing a lion or bear pacing back and forth in its concrete and steel cage. They are overpowered by the odor of urine on the cage floor, upon which the unfortunate animal must walk and sleep. The lion is certainly not happy in its tight, uncomfortable quarters. There must be a better way to keep and display animals, the visitors must wonder.

There *is* a better way. It is the outdoor zoo, where the animals are free to roam about in large, natural areas. The animals are healthier, happier, and much more interesting from a viewing standpoint to zoo visitors. This is the new concept in animal keeping and it is a good one. All kinds of animals are simply turned loose in large fields that are dotted with trees and rocky outcrops. The animals are segregated according to their compatibility. Naturally, you cannot keep lions in with antelope. You must study the natural history of the animals by visiting game parks in Africa, as well as those already established in our own country. In the open-air zoo, the animals are nearly as free as they were in the jungles. They have room to run and stretch their muscles, instead of pacing back and forth in a cage. Keeping and maintaining them is far easier and less costly, particularly in the South where climate is less of a problem. People view the animals in their own car, or by boat, minibus, monorail, or motorcade.

All kinds of animals are turned loose, even lions, gorillas, chimps, and baboons. So far no one has been eaten. In fact outdoor zoos have a better safety record than steel-cage zoos. Even with lions, bears, and other dangerous animals roaming free in large, outdoor enclosures, people are far less likely to get mauled, because they come in less contact with the animals than if the animals were held in a cage a few feet away. Nearly every year, someone is mauled or killed by a caged animal.

Busch Gardens in Tampa, Florida, keeps all kinds of animals, as well as thousands of birds of every description, outdoors in a veldt. You can ride across the veldt by monorail and see them all in perfect comfort and safety. Then there is Lion Country at Palm Beach, Florida, where hundreds of lions roam free. You can drive right in among them in your own car, in perfect safety. The lions sit under the shade of the trees and look at you as you drive by. You are as safe from them as you would be if they were behind steel

Little Spunky, one of the many cats in the author's household.

The ethereal beauty of a cypress swamp is more exquisite than any artist's painting. To visit such an area in a motorless boat or canoe is a journey into a fairyland of enchantment.

Two of the world's rarest and most prized animals are these fine Indian rhino at Crandon Park Zoo. Unfortunately, at present there is no way to see or photograph them except through this complex of heavy pipes, but when the new zoo becomes a reality, the author hopes that they will be placed in a large outdoor hammock with at least several acres just for them.

Coconut palms at Crandon Park Zoo. If the author had his way, he would allow the caged monkeys freedom to climb the trees. A smooth, ten-foot-high concrete wall around a grove of palm would keep the monkeys from climbing out. There would be no need for cage cleaning.

The Crandon Park Zoo has a variety of palm trees and tropical growth. When the zoo moves to its anticipated mainland location, perhaps the present zoo will be maintained as a botanical garden and aviary.

Crandon Park Zoo is in a tropical setting among swaying palm trees and ocean breezes, a most beautiful area for a zoo, except that hurricanes pose a constant threat to the animals.

The ocean is just a few hundred feet from Crandon Park Zoo. One can picnic and bathe at the beach after a stroll through the zoo. But it is the nearness to the mighty ocean that has aroused the need to move the zoo inland. During hurricanes, this area is flooded.

A beautiful lagoon inhabited by hundreds of birds at Crandon Park Zoo. A walk through the park is a refreshing experience, for the lush tropical growth adds to the beauty of the zoo.

bars (so long as you stay in your car). It is a fact that the lions look more natural out in the brush than they would in their concrete jails, and the keeping of them is greatly simplified. There is no cage cleaning required. The animals are simply out to pasture and all that has to be done in the way of maintenance is an inspection of the fences and feeding.

Even feeding is greatly simplified, especially in the case of giraffes, zebras, rhino, and antelope. These are all grazing animals, so they simply eat grass or shrubs like so many cattle. Supplementary feeding is all that is necessary. Again, no cage cleaning is required. Fewer zoo attendants are needed in the outdoor zoo than in the conventional zoo. There is virtually no need for cage cleaners, feeding is greatly simplified, the animals are healthier with far fewer diseases. This is the modern concept of a zoo. It is being adapted by many of the present zoos, but unfortunately, since most of the zoos are located in the heart of the city with scant acreage, they can do little more than plant a few trees or enlarge their cages to simulate the outdoors. This is a step in the right direction, but it is merely a treatment for a problem and not the cure. The cure is to move out of town to the very outskirts of the city. Keep the present zoo as a holding or staging area with limited viewing of the animals by the public, but move the zoo to large quarters. Do not settle for a few hundred acres for the new zoo. This is precisely what is wrong with the present zoos. They just do not have enough land to properly keep and display wild animals under modern concepts where the animals live and roam free, as they should. Purchase at least two or three square miles of wild, undeveloped land for the new zoo and keep the land in as near its original condition as possible, except to enhance it with additional trees, ponds, or wooded areas. You may find that you can buy several square miles of wild land out in the country for the same price that you would pay for a hundred acres in the middle of a

A jungle scene at Crandon Park Zoo makes a spectacular setting for a wildlife exhibit. This most tropical of United States zoos is a combination of the old and the modern zoo. The new proposed zoo could be one of the finest in the world.

A view across one of the lagoons at Crandon Park Zoo. Note the numerous water birds and beautiful trees.

city. Zoos belong out in the country. It is no problem to run buses out into the country for a zoo, and people associate animals with the wild outdoors. It is the perfect place for animals and the zoo.

Is the outdoor zoo more expensive to construct than the city zoo? No! It is much less expensive. Instead of all the concrete and steel prisons of the old-fashioned zoo, all that is needed is a good fence or a water-filled moat. The basic equipment needed to make an outdoor zoo is a ditch digger and a fence-erection crew. You simply dig a trench four feet deep and a foot wide and fill it with concrete. Then erect a ten foot chain-link fence on top of it. The imbedded concrete keeps animals from digging in or out, and the ten-foot-high fence will confine about ninety percent of all animals commonly displayed in zoos. As soon as you have your fence completed, you can start putting animals in your zoo. You do not have to build houses for them if you live in the southernmost part of the country. They can sleep outside in the wooded hammocks, just as they do in the wild. Of course you have to use discretion

Beautiful tropical growth at Crandon Park Zoo makes an ideal setting for this water bird exhibit.

A big step in the right direction is the new elephant compound at Crandon Park Zoo. This is a fine exhibit, with water and open space for the animals.

as to which animals can be turned loose, which will live together, etc. As I mentioned above, you cannot keep lions and antelope in the same outdoor enclosure. You will have to build more than one enclosure, or partition off the one you have, keeping the predators away from the grazers. But this is an easy zoo to build, particularly in the warmer parts of the country.

I have visited a number of these outdoor zoos and am convinced that they are the only really practical zoos. The animals are displayed in a natural setting, far surpassing any concrete cage, no matter how spacious. I visited most of the big-game preserves in Africa recently, from Kruger Park to Nairobi, and came back even more convinced that the natural way is the best of all. Kruger of course is a huge area, but much of it is now fenced to keep the animals away from the farms, for their own protection. We could have a dozen Kruger Parks in our own country for just the cost of one spaceshot or one napalm raid in Vietnam. In Livingstone, Zambia, I visited the Livingstone Game Park. This is a small preserve, just two square miles in area. A small bush and veldt area was simply fenced off to make the preserve, which in reality is an outdoor zoo. Then a number of giraffes, impalas, zebras, rhino, and other animals were turned loose inside. You drive through the area in a Landrover or a small minibus and the animals are all about you. They become so accustomed to cars that they will not even get out of your way. There are buffalo, waterbuck, wart hogs, and many other animals, all roaming free. There is a winding road that goes all through the park and is situated so that you cannot see the fence at all. We spent an entire afternoon in the preserve seeing all kinds of animals in their natural state. I think that the only zoo personnel in the entire place was the one guard who opened the gate. A regular zoo would have required fifty to several hundred employees to clean cages, feed the animals, and maintain the animals in a zoo "prison," and even with all this the animals would not be as healthy, as happy, or as interesting to view as those which are kept in a wide open zoo.

No visit to Crandon Park Zoo is complete for a youngster without sitting inside the huge mouth of the "hippo" at the children's zoo.

Does it not make more sense to build an open-air zoo? I think it does. I would like to see one built in Florida, South Texas, or California, not on several square miles, but on one thousand square miles. It sounds big and it is, but as I mentioned above, it would cost no more than a single bombing raid on Vietnam or no more than a single space launching, yet its benefit to America could hardly be measured. In such an area, we could turn loose an entire herd of elephants. All three major species of rhino, the white, black, and Indian, could be turned loose, where they could reproduce and someday be returned to their homeland to restock areas in which they once lived. Perhaps even a few of the rarer rhino now on the very verge of extinction could get a second start in such a preserve. Such a project would cost a billion dollars, but the money would flow back in almost as fast as the place was built. The smaller attractions along this line are a tremendous success. Who would not pay five or ten dollars for a safari ride through a jungle area complete with hippos and elephants? It cost me two thousand dollars for my trip to Africa. I think the public would buy the above project, and quickly!

An open-air zoo is practical. It really works and it is less expensive both to build and to operate. Some zoos are already working on it. San Diego's famous zoo is now setting up a huge wild-animal compound of three thousand acres on the open-air principle, and it is hoped that all zoos will get on the bandwagon. While it is true that the southern climates make the open-air zoo more practical, the same type of animal display can be set up in the cold climates. Wooded land with large rock caves and ponds is available up North too, and a great deal can be accomplished, even in cold climates, to better the living conditions of our wild animals held in captivity. In the North, of course, a heated house would be needed for most animals in the winter, but for six months of the year or longer, the animals could roam free. The Catskill Game Farm in New York and the Alberta Game Farm in Canada both keep all kinds of animals, from cheetahs to rhino, in giant, outdoor enclosures, not to mention Lions of Longleat in England, where the lions are left outdoors all winter long. They have shelters into which they can retreat if it gets too cold for them, but it is reported that they love the snow.

The outdoor zoo is *in,* and let us hope that the old-fashioned zoo, *without* its steel bars and smelly cages, will move to the country and give its animals their freedom. We owe it to the animals for the joy that they bring us. Zoos are our most popular outdoor recreation, according to statistics. Nearly everyone, young and old, likes to visit the zoo. But most everyone would rather see happy animals displayed in natural hammock or veldt or splashing about a large pond.

Bring up the subject of an outdoor zoo at your local zoological society and write to the press

A prairie dog living in a natural, tunneled area at the St. Louis Zoo. The trend in many zoos is to keep the animals in as natural a setting as possible.

about it. More than forty years ago, the Zoological Society of London decided to make just such a move. They decided that the animals at the zoo needed more room, so they purchased an old farm with several hundred acres and converted it into the beautiful Whipsnade Zoo, one of the world's finest zoos, where the animals roam free in huge outdoor paddocks. They moved their zoo out into the country, which is really where the zoo belongs. Now they have one of the world's most interesting zoos. At present, they have one of the largest herds of white rhino outside of Africa, with twenty-two of the giant beasts roaming free on a large tract of outdoor land. There is little doubt that the species will reproduce well in their roomy enclosure. Whipsnade is famous for its animal births. Already some rhino have been born there. Perhaps someday a few small herds of this gentle animal can be released in the southernmost national parks of America. It would be quite an addition to the fauna of the parks, and there is little danger of them multiplying like sparrows or rabbits. London was one of the first cities to move its zoo to the country to large acreage. It was a wise decision.

In Florida, one can see many animals, birds, and reptiles living in the healthful outdoors at Ross Allen's Reptile Exhibit at Silver Springs. Here may be found a tremendous collection of alligators, crocodiles, and, of course, snakes. Rattlesnakes, moccasins, and other poisonous snakes are on exhibit. Some of the largest captive alligators in the world may be seen here. This is one of the most famous wildlife exhibits in the state and carries a wide variety of specimens, including many mammals and birds.

There are also a great many of the so-called roadside zoos throughout Florida, as well as in all of the other states. Some are superb, with large paddocks and natural enclosures for the livestock, but some are pitiful. It is the latter that have given rise to the complaints of some vocal animal lovers who have made a blanket condemnation of *all* roadside zoos, not just those which are bad. This, of course, is not fair, and it is hoped that these same writers who have castigated all private zoos will realize the harm that they have done to the legitimate animal exhibits and will write articles about the good roadside zoos. They should do this if they have any sense of fairness. I have seen some roadside zoos that were quite superb, keeping their animals as well or better than many of the larger zoos. I feel that there is a tendency for some large zoos to monopolize the animal trade completely and to keep it among their close-knit society. This is also unfair and should be reversed, for it is really discrimination where the smaller zoo or animal exhibit is looked upon as inferior and forced to comply with expensive laws designed by and for the larger exhibits.

Zoos are necessary, not only for the display of animals, for which they were originally founded, but for the newer cause, which is the protection of our threatened species. Here, the zoos with their steady source of income, nontax status, and import priorities over private zoos, can make a real contribution to conservation. They can bring in hoofed stock for display, and then sell or trade the offspring to private zoos or animal breeders so that, with some endangered species, it will be possible to build up a reserve stock to insure survival of the species should some natural catastrophe destroy the remaining animals in their native land.

Not only do we need zoos, both public and private, but we need many more of them. As the threat of dwindling wildlife throughout the world becomes more real, the animals that have had their habitat destroyed by bomb or plow will have no place to go. Many will perish or be slaughtered by the soldier or the farmer. We need more zoos on a worldwide basis to offer maximum protection to our dwindling wildlife, for it would be a very dull world without animals.

But we need to modernize our zoos. Many of them are still living in the last century, so far as keeping and confining animals are concerned. Despite their know-how, some zoos do not confine their animals under optimum conditions. I have seen capybaras and tapirs, both water-loving animals, kept in dry cages at a public zoo. I have seen sun-loving lizards in a deeply shaded cage where the unfortunate reptiles would never have an opportunity to bask in the sun (which is vital to their health). I cannot help but feel pity for bears, lions, and tigers confined to a small cage for life. It is pitiful to see full-grown deer in a cage thirty feet long, with half a dozen animals crowded together for years on end, with never a chance to run or exercise. I have seen rats so plentiful in a large northern zoo that they would descend upon the food trays of deer and antelope by the hundreds, even biting the deer and undoubtedly eating the bulk of the food. And I have seen monkeys and chimps in steel-bar cages so small that the poor creatures could hardly take more than a few steps. All this and more exists today in which are considered modern zoos. Yet these very same public zoos are sometimes the first to criticize the private zoos. They are adamantly against the keeping of all exotic animals by anyone except themselves. I recently saw two adult marmosets at a public zoo, in a cage scarcely eighteen inches wide by less than three feet tall—this for an animal accustomed to great leaps through the treetops. One can easily find many examples of improper animal care in most zoos.

I am not pointing this out to find fault with the zoos, for it is understandable that public zoos in particular sometimes are very restricted in funds. In addition, politics play a part, so that a zoo director may have to put up with incompetent help that he cannot dismiss lest he jeopardize his own position. But it should be pointed out that before a zoo official condemns others for keeping animals improperly, he should make the rounds of his own zoo and examine every animal there to see that it is properly housed. He may find his own zoo guilty of the very charges that he has inflicted upon the private animal keeper.

There is also jealousy to be considered, as well

Giraffes at the San Diego Zoo Wild Animal Park. This is the new open-air concept in zoos, and is a marvelous way to keep animals. Courtesy San Diego Zoo Photo. By F. D. Schmidt.

as prestige. After all, a zoo may not look kindly upon another animal exhibit opening in the same area for fear that the new exhibit may steal a little of its thunder—particularly if the new attraction keeps and displays the animals in a beautiful natural setting that puts the zoo to shame. This is a human reaction, and rather than fight each other, the two exhibits could and should work together. They could help each other in many ways and both would end up with better places than before. In our days of growing population and expanding cities, one zoo is usually not enough for a community. In fact, most large cities could easily afford and benefit by as many as half a dozen zoos, particularly if the zoos are the open-air type. The educational benefits to the schools alone would more than justify the costs of such zoos, not to mention the aesthetic aspects to the general public, which flocks to the zoos in ever-increasing numbers.

Thus, not only do we need zoos, we need more of them, for a day at the zoo can be one of relaxation away from the smoke and grime of the city streets, a communing with nature, and an enjoyment of the wild animals from the far corners of the earth.

The San Diego Zoo is leading the nation with its huge, eighteen-hundred-acre Wild Animal Park. Here, a herd of white-bearded gnu roam freely in a natural area, much as they would in Africa. Courtesy San Diego Zoo Photo. By F. D. Schmidt.

Of course it must be pointed out that many zoos are already exhibiting many of their animals in barless cages and large outdoor moats, where the animals are kept under near ideal conditions. One could hardly find fault with the beautiful Detroit Zoo, which is one of the finest zoos I have visited. At one time Ralph Bowden and I made a complete, professional movie of the entire zoo there, which at the time was under the direction of the late Frank McInnis. It is certainly a fine, well-kept zoo and a natural park area of impeccable quality. Nor could one find fault with the San Diego Zoo with its healthful outdoor cages where bars are the exception rather than the rule. The giant, walk through birdcages with huge trees and shrubs and hundreds of exotic birds flying about is one of the wonders of our modern age. The Oklahoma Zoo with its huge

Waterbuck and sarus crane living wild and free in the fabulous new Wild Animal Park at San Diego Zoo. Not only do the animals appear more interesting in this natural area, but they are probably much happier and healthier than if confined to a small cage. The terrain even looks like the African bush. Courtesy San Diego Zoo Photo. By F. D. Schmidt.

outdoor paddocks and its record of breeding exotic antelope certainly deserves commendation, as do all of the zoos that are making a real effort to better the conditions under which their animals are kept. But it is hoped that the trend to even larger outdoor paddocks and preserves will continue. San Diego has apparently taken the lead with its huge eighteen-hundred- to three-thousand-acre animal park and already has released cheetahs and other animals into the area. Miami could be next with its envisioned Super Zoo, which will be planned on six hundred or more acres by the Florida Zoological Society and Crandon Park Zoo. I have loaned them some of my own material for possible use in the project. Other zoos perhaps are planning similar projects. It would not be surprising to see the very progressive New York Zoological Society come up with something that will be ahead of them all. They are also making great strides with animal exhibits at their local zoo, with veldtlike antelope and lion exhibits and even a rainforest—a remarkable achievement for such a northern area.

Everyone likes to be proud of his zoo, and you can speed up the modernization of your zoo to new, outdoor, natural exhibits by bringing up the matter each month at your local zoological society. You could make a project of modernizing one exhibit at a time, for a major overhaul of a city zoo is a project of such magnitude that it would almost defy possibility. Or, if your town or city is planning a new zoo, make certain that it is done with modern concepts of large open-air moats and hammocks, and reserve the bars only for human prisoners.

Let your imagination run wild. For a small in-city zoo, picture a small wooded plot of land with a twenty- or thirty-foot wall around it and a glass ceiling. Trees and shrubs would grow naturally from the ground and the whole thing would be both air-conditioned and heated to cut out city noise and grime. Inside, exotic birds, monkeys, ocelots, and smaller species of animals and reptiles would cavort about the trees and ponds in their respective enclosures. You would view them by walking through a glass-lined walkway with open viewing for some species. Lush tropical vegetation would absorb animal odors. Overhead sprinklers would keep delicate orchids and plants moist. There would be no flies, no filth lying on concrete cage floors. You would be in a miniature garden of paradise. A modest admission charge would finance and maintain the zoo. Large cities could have several, even dozens of such places, right in the heart of town. What a marvelous place for a midmorning coffee break.

Of course for the big animals, and the big zoo, it should be far out from town where there is sufficient land. Nearly everyone has an auto today, so getting there is no problem. There are also buses. With the big zoo, you can really let your imagination wander. Read a great deal, visit other zoos, and try to outdo them all with the most beautiful and natural zoo of all. Try to duplicate the wilderness, the jungle. Do not let anyone tell you it cannot be done, for today, little is really impossible. "Get an idea and follow it through. You'll be amazed at what you can do." A high school lecturer told this to our class thirty years ago and I have found it to be true.

10

Open-Air Zoos and Exhibits Already Available

In our own United States, the open-air veldt type of exhibit is being seen more and more. A new attraction called the World of Animals opened just outside of Dallas, Texas, under the auspices of Brian Hunt, the famous animal importer. It has over five hundred acres of land with fifteen hundred animals and birds, and the animals are viewed by driving through the preserve in your own car. The project is located near Mesquite, Texas, east of Dallas, and is designed to protect endangered species of animals that are not being given proper protection in their homeland.

Visitors to the park drive along six miles of simulated African plain, going through nine separate compounds, each featuring different species of exotic beasts. No one is allowed to roll down the car windows or drive through in a convertible. Rental sedans are available at the entrance.

As a security measure, armed game rangers are on duty in each compound 24 hours a day.

Since the animals are not shipped away during the winter months, the park must be maintained all year long, and therefore is open to the public every day of the year, from 9 A.M. to 5:30 P.M. in fall and winter, and until dusk in spring and summer.

Restaurants and picnic facilities are provided, and an authentic African Curio Shoppe offers a wide variety of unusual gifts and souvenirs.

Safari Village is a favorite attraction of young visitors to the park, since in this walk-through area, they are allowed to touch and pet young wildlife. During the school term, the park has its largest draw on weekends.

World of Animals is the largest attraction of its kind in the country and has gathered together more unusual animals than can be found anywhere in the world. More than $100,000 will be spent each year to feed the animals. A lion eats 12 pounds of meat a day at maturity, and a fully grown elephant eats approximately 350 pounds of hay a day, as examples of why the food bill is so high.

In addition to being a tourist attraction, the park will also serve as an animal survival center, offering many endangered species opportunities to thrive and reproduce, which no longer exists in their original habitats.

Additional unusual attractions are planned for the World of Animals in the near future, such as an African-motif restaurant, club, and four or five elevated guest cottages. Along with driving through the park, visitors will also soon be able to take a boat and a train safari through World of Animals.

Animal Management Services Inc., which manages World of Animals, has recently opened several other game preserves including Animal World in Rocky Point, outside of Providence, Rhode Island; African Safari in Port Clinton, Ohio; and Bob-Lo Island in the lower Detroit River.*

Africa, U.S.A.

There is an Africa, U.S.A., in California, where many of the animal movie sequences are filmed, such as in the movie *Daktari*. The animals here are given complete freedom of the lands. There are probably other, similar open-air animal exhibits in California and in other states. Unfortunately newspapers sometimes do not carry information about them, since they tend to promote tourist attractions in their own, immediate area. There could be a huge wild-animal attraction in

* Text courtesy World of Animals.

Curiosity is a common trait in man and beast as these dromedaries demonstrate by venturing close to get a good look at patrons of World of Animals. Visitors are advised to keep their car windows rolled up, thus avoiding any possibility of a friendly nudge or nip. Courtesy World of Animals.

your next county and you would hear little about it unless it were advertised in your own newspaper or on television. The newspapers just do not consider such attractions news unless the project has a good public relations department and can get stories to the press.

These baby Asian elephants, inhabitants of World of Animals, have an appetite that is huge even by Texas standards. Although neither is half the size it will reach at maturity (fully grown, each will measure ten feet in height and weigh more than two tons), the pair manages to consume 250 pounds of hay and grain daily. Courtesy World of Animals.

Even though they are ferocious looking, these lions can be very friendly and walk right up to you or stand on your car while you are going through the lion compound at World of Animals, which contains seventy-five lions. Courtesy World of Animals.

Two white rhino, the largest of the rhino family, get a firsthand look at Brian Hunt, General Partner in World of Animals. Six white rhino roam freely in their own compound, as do one hundred other species that make up the world's largest collection of exotic beasts. Courtesy World of Animals.

From any angle, the African crowned crane, with its golden tiara and multihued feathers, is one of the most beautiful birds abounding at World of Animals. Royal in both name and bearing, it does, however, avoid those areas inhabited by another species known as the "king of beasts." Courtesy World of Animals.

Alton Freeman, world-famous animal expert, raises many exotic animals in his beautiful mountain preserve in North Carolina. Here he examines some fine llamas, a South American animal. The author visited his animal farm and saw herds of zebras, black bucks, antelope, and other animals thriving in the beautiful wild setting. Mr. Freeman owned and operated the Rare Bird Farm in Miami for many years. He has imported many large animals from all over the world.

At Catskill Game Farm, Catskill, New York, most of the animals are kept in large, outdoor paddocks. Visitors may walk right inside with many of the animals. It is a great treat for the children.

Beautiful tame deer, at Catskill Game Farm. They are especially selected for the public feeding and petting area. Once you walk inside, hundreds of animals rush at you, especially if you are holding a milk bottle with which to feed the younger animals.

The author's daughter feeds a llama at Catskill Game Farm. The animal farm has a tremendous variety of animals and birds, including white rhino, cheetahs, ostriches, and nearly all of the larger zoo animals. They all survive in the cold northern climate, and many reproduce. Winter quarters are provided, but in the warmer months the animals roam free in large, natural areas.

The spectacular South American rain forest at Monkey Jungle, just south of Miami, Florida. Here one may walk through wooded paths among dense trees and vines and watch for monkeys in the branches overhead.

Monkey Jungle

In Florida, there are a number of wildlife exhibits in various parts of the state, each offering its own specialized fauna and flora. One of the most famous is the Monkey Jungle, just south of Miami. It has a beautiful, natural rain forest where South American monkeys live totally out in the wild, climbing and swinging through immense trees just as they would in their native land. This is the ideal way to keep and display monkeys. They are much happier and healthier than they could ever be in a cage. Visitors view them by walking through a beautiful, wooded path with exotic jungle foliage. The monkeys climb through the trees high above, and it is a thrill to locate them in the overhead branches. There is a feeding station at the rain forest, and at feeding time, hundreds of monkeys rush from the trees to feed. Many, particularly the squirrel monkeys, carry young on their backs. Even the rare, red-faced ouakari has reproduced in the freedom of the forest. A trip to Monkey Jungle is a must for visitors to Miami.

A squirrel monkey, living wild and free at Monkey Jungle, holds its foot over a key word on a sign in the rain forest.

A mother squirrel monkey with its baby in the jungle area. Many young are born here in this beautiful, natural area.

Immense trees in the rain forest house a variety of monkeys that are free to roam in acres of beautiful, natural forest. This is the ideal way to keep and exhibit monkeys and other animals. Not only do they look better, but they live and reproduce better. Also, there are no cage cleaners or attendants needed, and feeding is simplified because the monkeys find much natural food in the forest.

The monkeys peer down at visitors from the branches of tall trees.

Hundreds of squirrel monkeys live free in the rain forest. At feeding time they flock to the special feeding area from the nearby trees. Many are carrying babies.

A squirrel monkey rests on a large stone in the freedom of the forested area in which hundreds of monkeys live as freely as they do in their South American home.

A squirrel monkey resting on a leaf in the rain forest.

A monkey swings on a vine among impressive tropical foliage.

The red ouakari monkey lives in the rain forest and has even reproduced there. This is a rare species, and the rearing of young in captivity is to be commended.

A pretty marmoset.

A rare golden marmoset accepts a live cricket for dinner. This colorful little animal is also on the verge of extinction, and several pairs are kept in large, tree-planted cages in an attempt to breed them to ensure survival of the species. Their natural habitat in Brazil is being ruthlessly destroyed, as it is for so many other animals, by "progress."

A marmoset emerges from the tall trees to eat a tidbit at feeding time in the rain forest. Many of these smallest of monkeys live in the forest area. You can hear them calling from the highest trees as you walk through the dense forest.

A large colony of Java monkeys has been established at Monkey Jungle. They swim in a large pool, and leap from high rocks to retrieve bananas and other food. Unlike most monkeys, this species loves water.

An African monkey with young. Frank Dumont, owner of Monkey Jungle, hopes to establish an outdoor colony of African monkeys at the exhibit.

Monkey Jungle has a breeding colony of rare African monkeys that it hopes to eventually release into a large, natural area with no bars or wire. The same is planned for the chimps and all other animals so that eventually no animal will be in a cage.

A large, open-air enclosure houses the graceful gibbons and orangutans with a minimum of restriction, giving them nearly complete freedom. It is far superior to any steel cage.

A full-grown male orangutan in its cave home. It is the most spectacular of the apes.

The orangutan moves about very little except at feeding time, when it slowly walks about its airy enclosure. It lives peacefully with the lightning fast gibbons that share its abode.

The full-grown male orangutan is one of the most grotesque of all animals. The high head dome and huge, fleshy appendages of the face give it an awesome appearance. In the past it was called the wild man from Borneo. *It is a gentle and intelligent creature.*

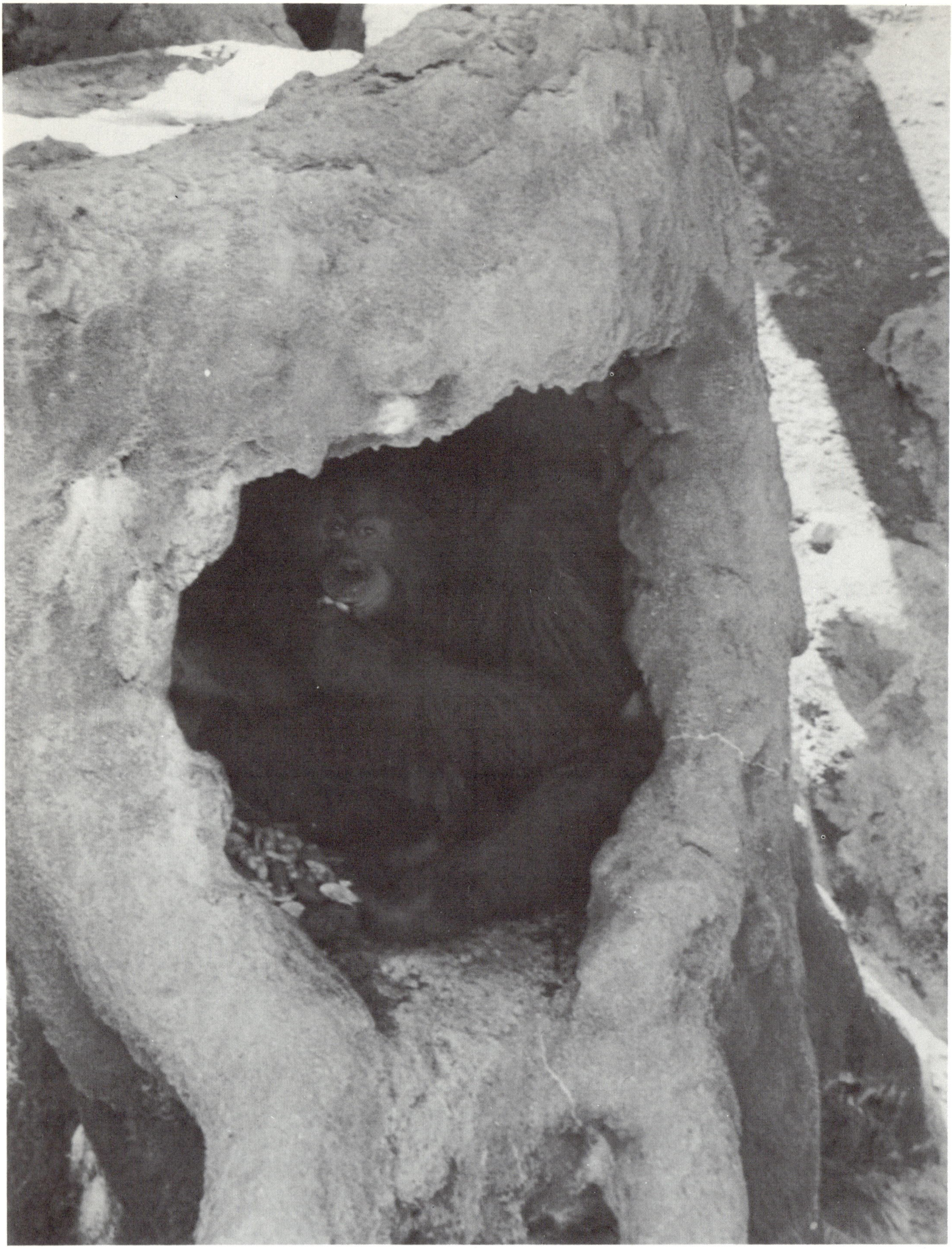

The female orangutan retreats to its own cave with pea-nuts and other food where it can eat in peace. Monkey Jungle hopes to breed this rare and endangered species.

Lions at Lion Country Safari in West Palm Beach, Florida. Here they live in huge outdoor enclosures, far superior to the tiny cages of the zoo.

The beautiful land in central Florida on which the author tried to build his African jungle. There are several hundred thousand wild and undeveloped acres in this one area alone. Enhanced with acacias and baobab trees, this would make an authentic jungle setting. No fences would ever be seen by the people on "safari."

Gorgeous flamingos parade in a tropical setting at Parrot Jungle in Miami, Florida.

The bright-red female orangutan climbs a concrete tree in its huge, spacious cage to get its food, which is placed in high areas. This encourages the normally lethargic animals to move about.

The ideal way to display and keep animals is the open-air compound, where the animals roam free in vast acreage. Visitors drive through the area in their car. Here at Lion Country Safari, visitors are greeted by African antelope and ostriches after passing through the gate.

Lion Country Safari

If you want to see lions, then you must visit Lion Country Safari near Palm Beach, Florida. This fabulous animal preserve has well over one hundred lions roaming free in a huge enclosure. You view them by driving through the area in your car, and you will not be eaten as long as you stay in your car with the windows rolled up. It is a marvelous way to see the king of beasts, wild and free instead of cooped up in a tiny cage. You will also see ostriches, giraffes, zebras, chimps, and rhino, as well as many other animals, in separate areas, of course. The animals have close to a thousand acres upon which to roam, yet you can come closer to them than you could at most zoos, and you could probably get better photos here of the lions than you would in Africa. At night, the lions are herded into a special cage for security purposes.

A similar project is located in England at Longleats Castle and is very popular in that country. Here, not only can you see all kinds of animals living free, but you can visit a real castle as well. It is reported that the lions stay out all winter, even though warm quarters are provided for them. Hippos, giraffes, baboons, and all kinds of animals live at the estate.

An ostrich looks in at some visitors at Lion Country Safari. How many zoos can offer such a rare treat?

A large male ostrich approaches a car. It is a powerful creature.

View of ostriches from car window.

Once inside the gates of Lion Country Safari, visitors follow the roads through vast areas of trees and brush, watching for animals along the way. Serengeti Plains is named after the African game preserve.

Then suddenly you see the lions up ahead. They are sitting under the trees for shade. They watch you approach in your car. You must keep your car windows tightly rolled up for safety, and you must drive slowly. If you wish to stop, you must pull to the right so that others can pass you.

The lions lie about in the open, resting contentedly.

The author saw this large pride of lions as he rounded a bend in the road. One of the park rangers came by and threw the lions a few bones. They all rushed toward the bones.

The lions milling around good naturedly. They are fed lightly during the day to make them move about for the visitors, but the choice food is saved for the evening meal when all of the lions are herded into an enclosure where they are locked up for the night, as a safety measure. The animals expect this and practically run to the stockade at the end of the day.

The lions settle down and chew on bones a few feet from the author's car. There are about one hundred lions in this exhibit, according to Bruce Grant, public relations official for Lion Country Safari. The lions have dozens, perhaps hundreds, of acres in which to roam. They are, of course, separated from the antelope and other animals.

One of the park wardens threw some chunks of beef around the author's car. Immediately he was surrounded by lions.

The lions settle down for a meal. They appear to be happy and content and act much like lions seen in Africa.

The lions have entire freedom in these huge tracts of veldt and brush. They must certainly be much happier here than they would be pacing back and forth in a steel cage.

You can usually get better photographs of lions at Lion Country Safari than you can get in Africa. This fellow walked right up to the author's car.

The author came across a small pond with lions sitting around it. He noticed a tiger among the lions. James Smith, a game ranger in a nearby zebra-striped truck, said that the tiger, named Janey, had been extremely ill and was given to the exhibit in hopes that it would recuperate. It did and is now a splendid animal. He said that it has chewed up numerous auto tires and must be watched constantly, unlike the lethargic lions.

Janey the tiger roams free with the lions at Lion Country Safari. It is stalking a chunk of meat here.

The lions approach to within a few feet of the car and show no hostility to the people inside. Even in Africa, lions rarely (if ever) bother people as long as the people stay inside their vehicle. It is foolish to take a chance with these animals.

When the attendant tosses some meat close by, all of the lions and the tiger spring into action. The tiger acts completely differently, crawling along the ground in a crouch as though ready to spring. The keeper, Jim Smith, calls out to it, and it obeys him. Perhaps it intends to chew on a car's tires.

Each animal grabs a chunk of meat or a bone and settles down again. The tiger retreats from the other animals. One of the male lions has befriended the tiger, which is a female, and it is hoped that they will mate to produce ligers *or* tiglons.

This African antelope has dozens of acres of land on which to roam. Crowned cranes are in the background.

Beautiful African crowned cranes feeding beside the road. They roam freely in a vast area at Lion Country Safari.

Zebras have dozens of acres in which to run. They keep the grass well eaten in the area. Large grazing areas are essential for hoofed stock in order to cut down considerably on the food bill.

At Lion Country Safari, visitors can obtain excellent close-up photographs of animals by driving within a few feet of them. The above photograph was taken with the normal lens of a camera. Note the healthy appearance of the animal.

Due to the difficulty in confining elephants, they are kept in a separate enclosure, but it is a large area, unlike the hundred-foot stalls of most zoos. Elephants can walk right through a chain-link fence, which of course would also let the other animals escape. They can be confined with a concrete ditch and moat, as shown here, or with railroad rails used as fence posts, hammered into the ground and strung with one-inch steel cable.

The attraction at Lions of Longleat in England was an immediate success. Millions of people came to see the animals living free, a milestone in wild animal keeping. World-famous animal expert Jimmy Chipperfield set up the animal preserve, which is one of the finest in the world. Lord Bath, Sixth Marquess of Bath, was the first to open his stately doors to the general public. Courtesy Lions of Longleat.

Contented lions at Lions of Longleat. The animals roam free in a vast, wooded enclosure and stay outdoors nearly all year round. They are provided with warm enclosures for winter but often romp about in the snow. There can be no question that lions, or any animals, are much happier in this ideal condition instead of in barred cages. Courtesy Lions of Longleat.

Lions of Longleat contains many other species besides lions. Here visitors mingle with numerous giraffes in an area devoid of lions. There are several species of giraffes at the exhibit, as well as zebras, cheetahs, chimps, and many exotic birds. Courtesy Lions of Longleat.

A band of baboons cavort about the cars at Lions of Longleat. Visitors drive through the preserve in their own car and see lions, hippos, giraffes, and most of Africa's other animals. The hippos swim in a moat around the castle at Longleat. Courtesy Lions of Longleat.

Tropical landscaping and an exotic fountain are focal points of Busch Gardens in Tampa, Florida. Towering above the gardens is the Anheuser-Busch brew hall, where Budweiser, Michelob, and Busch beers are produced. Above the seven-sided Hospitality House is the Stairway to the Stars, one of the longest outdoor escalators in the world. Some three million visitors tour Busch Gardens each year, making it one of Florida's most popular tourist attractions. Courtesy Anheuser-Busch.

Soaring eighty-six feet along the facade of the Anheuser-Busch brewery, the Stairway to the Stars takes visitors to a specially constructed observation deck on the roof of the Lager Cellars building. Courtesy Anheuser-Busch.

Busch Gardens

Busch Gardens in Tampa, Florida, is another paradise for wild animals. Huge herds of zebras, giraffes, gnu, even elephants, all roam free in a veldtlike setting. You view them from an overhead monorail, and it is like a trip to Africa. I have visited the exhibit many times. Richard Naeglie, who was director at the time, drove me among the animals in a Landrover and showed me how the lions could be confined if necessary. Many exotic animals have been raised there, which of course is a great conservation effort. Also, a great many spectacular and colorful birds have been reared in the gardens. Many of the birds fly free and build nests in the trees throughout the spacious grounds. I have seen the crowned pigeon calmly nesting just a few feet over my head in trees right by the walkway where thousands of people pass each day. It has been reported that Busch Gardens is now expanding into large acreage of immense tracts, some which will encompass several thousand acres. Perhaps they will build a real jungle in America, which has been the dream of many an animal lover, including myself. The famous author of *Animal Kitabu* and *Congo Kitabu,* Jean-Pierre Hallet, also dreams of building an authentic big-game preserve somewhere in America, and I wish him every success. His books are fabulous accounts of wildlife in the Congo and must reading for anyone interested in animals and wildlife.

Monorail cars pass in front of the Old Swiss House restaurant as they return from a Skyrail Safari through the Wild Animal Kingdom at Busch Gardens. Framing the cars in the foreground are representatives of the Wild Animal Kingdom and garden areas. Courtesy Anheuser-Busch.

Skyrail Safari winds its way around the northwest portion of Wild Animal Kingdom, where South African blesbok graze. Courtesy Anheuser-Busch.

Herds of wild antelope, buffalos, and deer roam a recreation area of their natural habitat at Busch Gardens. Courtesy Anheuser-Busch.

Herds of wild animals from Africa gather around "Lake Victoria" at Busch Gardens. Many species located in the Wild Animal Kingdom of the gardens are threatened with extinction in Africa and are protected at Busch Gardens in a re-creation of their natural habitat to encourage reproduction. Courtesy Anheuser-Busch.

A group of friendly giraffes tower above former Busch Gardens Zoological Director Richard G. Naegeli. The giraffes are among hundreds of African animals currently in residence at the Wild Animal Kingdom. Courtesy Anheuser-Busch.

Leo, a sleepy lion, passively watches the Skyrail Safari as it wends its way through the veldt. This healthy specimen is one of more than a dozen lions on display, including a pride of cubs born here recently. Courtesy Anheuser-Busch.

Passengers aboard the Skyrail Safari get a panoramic view of the 205-acre Wild Animal Kingdom from this point of the safari. Far in the background is the Old Swiss House. Courtesy Anheuser-Busch.

Bathsheba, a young cheetah at Busch Gardens, is very well fed by its keepers, but prowls its moated home as it would in the wilds of Africa. Courtesy Anheuser-Busch.

These cubs are not quite old enough to be interested in the affections of pretty Candy Shagaric, but rather, turn their attentions to a warm saucer of milk. The youngsters are only two of a pride of lion cubs born recently at Busch Gardens. They will make their home in the authentic veldt there. Courtesy Anheuser-Busch.

Spectacular performances by highly trained macaws and cockatoos are staged continuously throughout the day at Busch Gardens. Birds are taught to perform with almost human understanding. They slide down high wires, stand on their head, and perform many other feats to the delight of young and old alike. Courtesy Anheuser-Busch.

Two young visitors enjoy the view while playful macaws frolic in front of the Hospitality House, a focal point of Busch Gardens. The seven-sided building overlooks a fountain and lagoon where exotic waterfowl and large goldfish make their home. Courtesy Anheuser-Busch.

Busch Gardens' enchanting Storybook Lane, where legendary "little people" come to life. Here Rip van Winkle, Hansel and Gretel, Cinderella, Little Red Riding Hood, and others are characterized in carved figurines imported from Germany. Courtesy Anheuser-Busch.

Busch Gardens is truly a photographer's paradise. Whether you are shooting in color or black and white, hundreds of spectacular scenes are found at the gardens, such as the beautiful black swans, the graceful flamingos, and other rare birds. Courtesy Anheuser-Busch.

A huge giraffe in the spacious outdoor veldt at Busch Gardens.

This baby South African animal, a member of the antelope family, keeps a wary eye on the photographer. When fully grown, this little fellow will stand seven feet high at the shoulder, and his horns will extend three feet above his head. Courtesy Anheuser-Busch.

Graceful flamingos at Busch Gardens. The colorful birds are so contented in their outdoor home that they even built nests and reared their young. Note the white egg at the center of the picture.

One of Busch Gardens' orangutans appears to be a little timid. Perhaps he has never held hands with such a pretty girl before. Courtesy Anheuser-Busch.

"You go first. . . ." A chimpanzee shows this young gorilla how to get into trouble on Ape Island at Busch Gardens. Courtesy Anheuser-Busch.

The popular picture of a huge male gorilla furiously beating his chest and charging a native is wrong; the gorilla does not have this habit. He is shy and retiring, and does not attack anyone without reason. Hercules can be seen at Busch Gardens sunning himself and feeding on plant material. Courtesy Anheuser-Busch.

Huge white rhino at Busch Gardens. It is also called the square-lipped rhino, and is the most gentle of its species. Just a decade ago it was on the verge of extinction but has made a remarkable comeback through rigid protection.

Sleepy orangutans at Busch Gardens are content to listen to bedtime stories before their afternoon nap. Courtesy Anheuser-Busch.

Pretty Cheryl Johnson gets inquisitive looks from a herd of friendly giraffes in the Wild Animal Kingdom of Busch Gardens. Dressed in a giraffe outfit, Cheryl had no trouble making friends with these long-necked denizens of the African veldt. Courtesy Anheuser-Busch.

One of Busch Gardens' towering giraffes peeks around the corner of a support bar and into the eyes of surprised passengers aboard the monorail. Visitors are able to view and photograph rare animals at extremely close range from the monorail. Courtesy Anheuser-Busch.

Richard Naegeli brings a young African elephant to attention for a photograph. In back is the Old Swiss House restaurant, where visitors may dine while watching a spectacular panorama of wildlife.

The author's son, Paul, poses with a bevy of colorful macaws at Parrot Jungle.

Parrot Jungle

There is a Parrot Jungle in Miami, which specializes in parrots, macaws, and exotic birds. This is a beautiful attraction and again, as with the other attractions mentioned above, many of the beautiful birds are allowed to fly free rather than be confined to small cages. It is one of the most beautiful and well-kept attractions in the nation and should be of special interest to bird lovers, for its collection of colorful birds is unique.

Parrot Jungle in Miami, Florida, is among the most beautiful attractions in America featuring wildlife in a natural setting. Here gorgeous flamingos parade in a tropical setting. Colorful parrots and macaws fly free in the area and have made the jungle their home. There is a large collection of the world's most colorful birds at this attraction.

The Serpentarium

Miami also has one of the world's finest reptile exhibits, the Serpentarium. Here is found one of the largest collections of poisonous snakes in America, perhaps in the world. It is owned and operated by Bill Haast, who perhaps has the distinction of being bitten by poisonous snakes more than anyone else in the world. Mr. Haast extracts snake venom from cobras and other highly venomous snakes and has become world famous in his field. At one time, I leased a building at the Serpentarium for my saltwater aquarium business and had an opportunity to see Mr. Haast in action many times. He has "milked" poisonous snakes thousands, perhaps hundreds of thousands, of times, but it is always an exciting thing to watch, although frightening, to say the least. His pretty and charming wife, Clarita, sometimes assists him as he calmly extracts deadly venom from his charges. On Sundays, Bill Haast hauls out the giant king cobra and extracts its venom. This is a spectacle that probably cannot be seen anywhere else on earth. The huge cobra, which measures more than twelve feet, is tossed on the open ground, right in front of a group of startled tourists. Then Bill slowly approaches it, and using one hand to attract its attention, he deftly grasps it by the back of the head with the other. Then he proceeds to extract its venom. I have watched him do this a number of times and I must admire his courage. On one occasion, he was bitten by the giant cobra. Fortunately he survived, as he has survived well over a hundred other bites by deadly reptiles. I have often told Bill that I would never give him any competition with his deadly snakes.

The exhibit also has a number of lizards, crocodiles, and hundreds, perhaps thousands, of snakes, many displayed outdoors in beautiful, tropically planted enclosures, where the snakes are often found in the trees overhead. Recently, a new addition to the exhibit has made its appearance. Bill's daughter, Naia, now has a pet lion. It is just a cub but it is growing fast. I went to the Serpentarium especially to photograph it. I had seen Mrs. Haast at the postoffice and she told me about it, so I made a quick trip to the famous exhibit. I was inquiring about the cub at the gift counter when the little fellow came running out of the back room and bumped right into me. It was the first lion I had seen since my trip to Africa. Naia Haast takes the lion out for strolls on a heavy chain leash, but already the king of beasts is showing signs of strength. It is anyone's guess how long the strolls will continue. It is easy to see why Naia would want a baby lion. It is such a beautiful animal and so exotic. Besides, it is quite a departure from the usual snakes.

World-famous snake handler William E. Haast prepares to "milk" a snake for its venom at his exhibit, the Serpentarium in Miami.

The snake is removed from its cage with a snake hook while a group of tourists back away. Bill Haast bears the distinction of being bitten by more poisonous snakes than anyone else in the world. He has suffered over one hundred snake bites, including one from a king cobra.

A pretty assistant stands by while Bill Haast fills out a card. Meanwhile, the snake on the scale is slowly crawling out.

A death duel with a giant king cobra at the Serpentarium is held every Sunday by Bill Haast. The show begins when he arrives carrying a huge king cobra. He deftly flings it to the ground right in front of the visitors, who quickly back out of range.

Then he moves in front of the snake, studying its behavior.

Next, he settles to the ground next to the snake, moving one hand in front of its face to draw its attention. Note that the deadly snake has its mouth open, ready to bite.

Then he rises to his feet, keeping his hand moving. The snake also rises higher off of the ground.

The powerful reptile twists and squirms, trying to get away, but Bill hangs on with what must surely be one of the world's most powerful grips.

While the snake is watching his right hand, he then reaches in quickly and grabs the reptile firmly with his left hand. One miss at this point could be his last. In fact, he was bitten by a king cobra in one of his performances and is one of the few people to ever survive such a bite.

The finish comes when the reptile is held to the milking flask where the venom is extracted. It is a spectacular though chilling sight to see and probably cannot be seen anywhere else in the world. The author was headquartered at the Serpentarium at one time, using a large building that once had held thousands of poisonous snakes, but he always told Bill Haast that he would never give him any competition (with good reason).

Alligators in a natural setting at the Serpentarium.

A fearsome Nile crocodile at the Serpentarium. This fellow is over twelve feet in length.

A giant monitor lizard at the Serpentarium. Many of the lizards and snakes are kept in beautiful, outdoor enclosures of rounded concrete, with natural grass, rocks, and trees in the center. This is the ideal way to keep reptiles and is far superior to a wire cage.

Deadly water moccasins, one of Florida's poisonous snakes, one coiled and one uncoiled, on exhibit at the Serpentarium.

The monitor lizard will hiss loudly if approached too closely.

A pygmy rattler in one of the pits at the Serpentarium.

Rainbow Springs

"Rainbow Springs, situated on one of the world's largest freshwater springs, has taken advantage of this wonder of nature to provide visitors with a unique three-dimentional tour of the broad, crystal-clear Rainbow River.

Whether your view is from an underwater boat, a turn-of-the-century riverboat or a colorful river raft, the beauty is unsurpassed anywhere on earth. . . .

There are many other points of interest at Rainbow Springs—thousands of colorful birds, hundreds of thousands of plants, registered quarter horse farm, western-style rodeo and a fascinating aerial monorail.

But the sparkling Rainbow River, named for the multi-colored vegetation which reflects a rainbow-like pattern in the sun-light, is the centerpiece.

One of the most popular ways to take in this extravaganza of nature is aboard one of the specially-built, 24-passenger underwater boats. Picture-type windows beneath the waterline provide viewers—seated five feet below the water's surface in air-conditioned comfort—panorama of the underwater world.

Unlike the more commonly-used glass-bottom boats which limit the view of the passengers to directly under the boat, Rainbow Springs' underwater craft permit an unobstructed view of the mysterious underwater world for nearly 360 degrees.

Looking through the one-half inch thick glass at 50 species of fish and turtles and 37 varieties of marine plants, visitors feel like intruders in a strange world. Colorful fish swim right up to the window to peer in.

Included in this natural setting are largemouth bass, which weigh up to 20 pounds, and prehistoric leopard garfish, which are much the same in appearance today as they were centuries ago.

The Rainbow Queen riverboat provides a completely different view of the river. This authentic stern-wheeler, much like those that were used to settle Florida in the 1800's, treats visitors to a leisurely cruise along the surface of the historic river as it winds its way through a forest of trees and plants.

The third facet of the water-oriented tour is the River Raft Adventure. Like the Rainbow Queen, this ride is also a step back in time, to the days of Huck Finn and Tom Sawyer. A native guide ushers guests onto the rustic wooden river raft for a lazy trip through the shallows and bayous of the river.

From this unobtrusive vantage point, visitors can observe many varieties of birds and wildlife existing peacefully in an undisturbed natural habitat.

In this era of environmental concern, Rainbow Springs has displayed an attitude of ecological responsibility applauded throughout the state of Florida. . . .

Back in 1968, when the attraction began its last major expansion, plans had already been formulated to avoid disturbing the area's ecology.

Most notable example of this is the conscientious landscaping necessitated by the construction of the Forest Flite monorail.

Unlike the most common method of constructing a monorail, which is a direct line from one point to another, Rainbow Springs' Forest Flite winds strategically around trees and plants, so as not to disturb their natural habitat.

Moreover, the Forest Flite was built on hilly ground composed, in many cases, of solid rock. Consulting engineers said it would be impossible to set the steel stanchions, necessary for supporting the monorail, into the rock without blasting with dynamite.

Undaunted, Rainbow Springs' management undertook the job with conventional machinery. As an added precaution, a team of tree surgeons was employed to follow the construction workers the entire length of the monorail, repairing any damage to nearby trees which inadvertently occurred in the process.

While this was taking place, the land above the river was terraced to retard soil erosion and resulting pollution.

'Rainbow Springs shows that man and nature can live together,' said Billy Puryear, general manager. 'We are not super-conservative conservationists who say you cannot put a boat on a river or build a path through the woods because you destroy the ecology. Rather, we take the attitude that we can create a beautiful attraction that complements nature.'

'When we undertook the vast project of landscaping the attraction in 1967, we looked at each individual area with an eye toward making the best possible use of the natural beauty of the area and then enhancing that beauty.'

*The balance of nature has remained so constant that there has never been any need to spray insecticides at Rainbow Springs. 'We did exactly what we set out to do, ecology-wise,' Puryear concluded."**

---

* Reprinted courtesy Richard L. Needleman, Rainbow Springs, Dunellon, Florida.

The unspoiled natural beauty of Rainbow Springs is typified by this view of the historic Rainbow River. From the green adjoining the dining terrace, visitors can watch old-fashioned riverboats and unique underwater boats as they journey down the river. Springs that feed the river produce millions of gallons of crystal-clear water daily—enough to supply a large, modern city. Courtesy Rainbow Springs.

This talkative Rainbow Springs resident went out on a limb to chitchat with two comely visitors about the thousands of colorful birds that inhabit the eight-story aviary. Courtesy Rainbow Springs.

The unique Forest Flite monorail at Rainbow Springs carries passengers through the treetops at a leisurely pace, providing a breathtaking view of majestic waterfalls, colorful gardens, and the world's highest aviary. Courtesy Rainbow Springs.

Karen Jones lifts a page from Huckleberry Finn *and* Tom Sawyer *as she shoves off for a day of bayou and river adventure at Rainbow Springs. She is aboard the* Rainbow Raft, *which makes daily guided tours of the crystal-clear Rainbow River.* Courtesy Rainbow Springs.

The sixty-foot cascade of Rainbow Falls, highest in Florida, inspires a blithe forest beauty to this exuberant pose. The source of the falls is Rainbow Springs. Courtesy Rainbow Springs.

The Rainbow Queen, *a double-decked riverboat, stirs memories of Florida's early 1800s as it cruises leisurely down the historic Rainbow River.* Courtesy Rainbow Springs.

Jungle Larry's African Safari

"'The name of the game is conservation,' remarked TV's JUNGLE LARRY or in more scientific circles as Lawrence Erskin Tetzlaff, owner of the African Safari *near tropical Naples-on-the-Gulf of Mexico, Florida. He was introducing wildlife programs his SAFARI is producing right here in America to a group of Mid-West reporters on tour of his authentic jungle complex.*

Standing in the middle of his lush congo-like jungle, pointing out many of his rare vanishing species of animals and tropical trees, he said with great emotion, 'Let us be concerned with the conservation of the world's animals and plants . . . within one year 89 species of fish, birds and mammals have become extinct. 150 more are endangered. These are part of our common heritage. We must understand, protect and preserve them. Our entire staff is dedicated to preserving and breeding many of the world's rarest animals and trees.' Tetzlaff has shipped many of his rare offspring to zoological parks throughout the world.

JUNGLE LARRY started his wild animal career as a herpetologist (reptile expert) with the late Frank "bring 'em back alive" Buck. Buck was the superstar wild-animal trapper and exhibitor of the '20's, 30's and 40's. While Tetzlaff was with Mr. Buck in those later years he learned to handle everything from anteaters to zebras. Today many zoologists recognize JUNGLE LARRY as Frank Buck's successor. Bill Gordon, popular California radio, TV talk shows and renowned circus buff once said, 'All that Frank Buck and Clyde Beatty ever hoped to be, Jungle Larry is.'

One of the other highlights in Tetzlaff's career and one of his major turning points on the road to success was meeting his mate, SAFARI JANE, a devoted animal lover, trainer and photographer. Together they have gone to the remote fastnesses of the earth in search of rare animals. Their goal has been to capture pairs for their breeding programs in Naples.

The Tetzlaffs have even gone so far as to start producing new species of animals. One of their outstanding accomplishments has been the cross breeding of a Royal Bengal Tiger and an African Lioness. The result? Tiglons!!! Now at two years of age they appear to have lion heads and necks and tiger striped bodies. When they (two males and a female) are old enough to breed, they could produce: tiglons, tigers or lions. The female was born pure white and got her lion and tiger colors many hours later. Because a white gene is present an inbreeding program is planned to produce a white tiglon, tiger or lion. Millions have seen the tiglons on television. They have been featured twice on the NBC "Today Show" and over two and a half million people have seen them in person.

The setting for this unusual program to save some of the world's animals takes place in Naples, Florida down on the southwest coast of the Sunshine State. Here in scores of acres lie 54 years of effort on the part of three men with a dream. In 1917, Dr. Henry Nehrling believed that subtropical Naples could be a jungle paradise with its perfect climate. The doctor planted thousands of tropical trees and orchids from Africa, Asia, Islands of the South Pacific and South America. His dream came true. The jungle vegetation flourished in Naples with its gentle Yucatan breezes. After his passing he has been recognized as one of the world's great horticulturists. Dr. Witmer Stone, of the Academy of Natural Sciences of Philadelphia, said of him: 'A typical German professor of the old school, of courtly manner and enthusiastically absorbed in his work, he made a host of warm friends and was pleased as a child when visitors admired his garden.' After his passing the gardens were unattended from 1929 to 1951. The garden turned into a miniature congo-jungle.

In the latter part of 1951 the gardens were 'discovered' by Julius Fleischmann, the famed Cincinnati philanthropist. He purchased the entire area of land that contained Nehrling's labors. Gradually jungle trails were cut thru the wilderness and opened to the public as Caribbean Gardens. In the years that followed Mr. Fleischmann added many orchids and rare flowering trees to the botanical paradise. Then to add animation to his beauty spot he traveled the world in search of thousands of rare birds to enhance the complex. His specialty was waterfowl. In a few short years the collection grew to the second largest waterfowl collection in the world and the honor of the finest free flying water bird display. Some of the most colorful birds on earth are to be found in the garden.

One of Mr. Fleischmann's dreams was to have the gardens eventually become one of the outstanding botanical-zoological parks. After Mr.

Fleischmann's passing in late 1968, his son Charles asked Lawrence Tetzlaff if he would take over the project and fulfill his father's dream. Tetzlaff did accept.

In 1969 large ecological areas were fenced in with 15 foot high heavy gauge chain link. The most natural conditions were presented to apes, antelopes, birds, leopards, lions, tigers and scores of seldom seen animals.

A recent merger with the INTERNATIONAL ZOOLOGICAL SOCIETY and JUNGLE LARRY'S AFRICAN SAFARI assures the public and the world of zoology that they will witness the truest reproduction of the congo. They are calling it 'African Safari' and it is! The land of happy animals—no poverty, no pollution and no poachers!

*Another bonus awaits Naples, Florida visitors, a seven mile beautiful coconut palm lined non-commercial beach. A taste of Africa and a sample of the old unspoiled Florida."**

*Reprinted courtesy Jungle Larry's "African Safari," Naples, Florida.

Jungle Larry and pygmy hippo Fine Boy from West Africa, at Jungle Larry's African Safari. Courtesy Jungle Larry's African Safari.

"Safari Jane" Tetzlaff gets a morning workout with a 350-pound tiglon, *product of the cross-breeding of a royal Bengal tiger and an African lioness. Safari Jane has spent the last fifteen years traveling in Africa and South America, capturing and training animals.* Courtesy Jungle Larry's African Safari.

Feeding the man-eating fish. Courtesy Jungle Larry's African Safari.

Friendly Toco the toucan will pose with visitors at Jungle Larry's African Safari. Toucans have a light beak, despite its ponderous, Durante-like appearance. The beak is not solid, but honeycombed, thus achieving surprising lightness. If it were not—what a load to carry! Courtesy Jungle Larry's African Safari.

Native village at Jungle Larry's exhibit. Courtesy Jungle Larry's African Safari.

Black-necked screamers from Brazil living in a natural setting. Courtesy Jungle Larry's African Safari.

Beauties and the beasts. Pretty girls hold adorable little lion cubs. Courtesy Jungle Larry's African Safari.

11

Build a Real Jungle in America

Did you know that there is a herd of zebras living out in the wild right here in the United States? There is, and in fact there are also eland antelope and other species all living on a beautiful island in Northern Florida. Many years ago they were released there and today they are still thriving (or were at this writing). The island is under study right now for some sort of national park, and the Department of the Interior has tentative plans to possibly remove the animals that are already living there and place them in zoos. What a horrible thought. Here we have exotic animals already living out in the wild on an island, where they cannot possibly do anyone any harm, and our Fish and Wildlife Service wants to put them in a zoo. Let us hope they do not do it. (I have written to them about it.)

We need just the reverse action. We need to take the animals from the zoos and release them out in the wilds. This does not apply to all animals, of course, for we could not very well turn tigers and lions loose, but the hoofed stock, such as the zebras, antelope, giraffes, monkeys, and birds, could do well out in the wilds and would add great interest to our countryside. I think a flock of ostriches running across the open land in the Southwest would be quite a pretty sight, as would a herd of the beautiful gemsbok from South Africa. Already one state, New Mexico, has imported these exotic animals for possible liberation out into the wild areas of their state. New Mexico turned loose a number of Barbary sheep in the mountain areas of the state, and they did so well that now limited hunting of the sheep is allowed. Now the New Mexican government is interested in gemsbok and greater kudu as well as impala and other exotic animals from Africa. If it can give these imports a second chance for survival, it will be doing our wild creatures a good turn, for already many interesting species of animals are in danger of extinction. I think such marvelous creatures as the cheetah would survive in such areas as the Big Bend National Park, where the deer are overpopulated and need some predators to keep them in check. Areas like the Big Thicket in Texas, which I have personally visited, would support a huge population of strange and wonderful African animals such as forest elephants, a wide variety of monkeys, baboons, exotic antelope, giraffes, ostriches, and possibly even a few rhino. In fact I made a trip to the area for the precise purpose of possibly establishing an African big-game preserve there, by personal invitation of Mayor Demsey Henley of Liberty City, Texas. It is a fabulous wooded area.

Other parts of the United States in the warm southern areas could provide a haven for some of the big game animals now threatened with extinction. I would not doubt that there are vast areas in Southern Texas and California where even elephants, rhino, giraffes, and most of the big-game animals of Africa could be turned loose, for the preservation of the species. Our National Park Service could conduct guided tours through the areas in Landrovers, as is done in Africa, so that the public could view the animals out in the wild. It would cost a few million to do it, or even half a billion if it were to be done properly, but this is a drop in the bucket for a country that can spend thirty billion dollars a year on an unpopular war. With the expense of a few million dollars, huge lakes could be built in what are now arid

Typical waterway that could be constructed in a big-game preserve. In Florida, such canals are easily dug. The fill at the left could either be left where it is, for flood control (in which case fences could be placed behind it, out of sight), or it could be removed (leaving only the waterway). A series of ponds and lakes could easily be dug, connected by the waterways. These could contain herds of hippos and crocodiles and numerous islands housing monkeys, gorillas, gibbons, and chimps. In some instances, the fill could be squared off as a wall to divide certain species.

areas, and irrigation canals could be dug that would completely encircle the areas. These canals could be dug in winding, irregular patterns so that they would give a picturesque ride through the wild areas by riverboats. Along the banks of the canals and lakes, exotic banyan trees and wild fruit trees, as well as edible and useful shrubbery, could be planted. In the warmer climates, baobab, acacia, and other African trees could be planted. This would add authenticity to African areas, which would be of educational as well as aesthetic value. In addition it would provide valuable food for various species of animals and birds. We could also introduce a great many exotic birds

Wild areas such as this in Southern Arizona could harbor exotic antelope from the desert areas of Africa. The gemsbok of Southwest Africa would be particularly suited to the area.

Impala at Kruger Park in South Africa. It is a thrilling sight to drive through the famous animal preserve and watch for animals in the dense forest. Building a Kruger Park in America could easily be done and would give people a chance to see Africa's wildlife roaming free in our own country. It is simply a matter of putting a heavy fence or a high wall around a large area of trees and brush, then stocking it with animals. Of course it could only be done in the southernmost areas of the country or in Central or South America.

from Africa that would further enhance our countryside. The vulturine guinea fowl, the African gray parrot, ground hornbills, secretary birds, and many species of highly colored birds would do well in our country. As mentioned before, if any of these species *did* become overabundant, modern man's killing power could easily exterminate them. As for the big-game animals, particularly the exotic antelope like the kudu, gemsbok, and eland, should these become too numerous, they could be killed off by allowing sportsmen to hunt them with special hunting licenses or by being hunted by the park rangers. The meat could be sold or distributed to the nation's hungry. I do not think that we should flood *all* of our wild lands with imported species. Areas such as the

Everglades National Park that are already rich in natural fauna should be preserved as they are. But there are vast and huge tracts of nearly useless land in Florida, Texas, and Southern California that could be developed into fantastic wildlife areas with the expenditure of a few million dollars and the imaginative use of a dragline, a bulldozer, and landscape architects. Vast areas in Southern Florida that are now nothing but flat, semi-arid wastelands that become mosquito beds in the summer and fire traps in the winter could be made into giant, deep lakes with surrounding islands and hills that could be planted with lush, tropical growth. Huge herds of hippos, crowned cranes, flamingos, waterbuck, impalas, baboons, patas monkeys, and chimpanzees—all would and could thrive in the area and would add to the interest and economy of the area. If the hippos became too plentiful, the excess could be cropped out and sold as hippo hams or meat. The flesh of hippos is said to be far superior to beef, so we would have a source of meat for the nation's poor as well as a gourmet treat for the nation's wealthy. If I had my way, I would make the entire state of Florida into one giant game preserve and let people come only to visit, but of course this is impossible. We can have some exotic wild areas in all of the southern states, and it is my belief that it would do far more good than harm and that the cost of a few million could not be better spent. We have a government that is always telling the people to stay home and see America, but millions of animal lovers want to see something besides an occasional deer or raccoon, which is about all the wildlife one can see at the present time. The northern climates will always have their deer and raccoons, and it is only in the southernmost part of our nation that any of the exotic African species could even survive because of the warm climate. Surely those purists who think that we should not bring in foreign game should realize this. After all, both the white man and the black man are imports, and I do not feel that we have improved our country much. In fact I think we have just about

Elephants at Amboseli Preserve in Africa. The author feels that a few herds of elephants would add interest to some of our southernmost national parks and would present a foothold in America for this spectacular beast. Each year, thousands of elephants are shot or die in game preserves in Africa because they have become too numerous for their limited boundaries.

Africa, U.S.A., in Boca Raton, Florida, was one of the first African jungles in the nation. The animals thrived and even reproduced in the beautiful natural setting. A trip through the area was almost like a trip to the African veldt.

Zebras at Africa, U.S.A. appear as natural as they do in their homeland. The attraction captured the imagination of the animal-minded public and drew enormous crowds, but now housing lots have taken over the valuable real estate. The animals were sold to the nation's zoos. Apparently there was not enough natural grazing land, making an enormous feed bill, which is why the author suggests a very large tract of land for a jungle project.

ruined it. We have turned the rivers into sewers, turned the woodlands into dumps, and are even polluting the mighty oceans. It is time now that we turn about and help the wild creatures that live on this planet with us. It would not destroy America to have a few herds of wild elephants running loose in some protected area, nor would a few small colonies of white rhino in the vast stretches of Southern Texas or California do

African buffalo and waterbuck at Treetops in Africa. The famous lodge is built high above the ground in a setting of large trees, and from its rooftop or windows a fantastic panorama of wildlife may be seen. The animals parade out from the forest and settle down to dig into the earth for salt, thirty feet below. We could have a similar spectacle in America by building a treetop lodge for animal viewing in a large wilderness area in the southernmost part of the country.

serious harm to the country. I think that it is time that we give some serious thought to this big-game business. It could mean big money to the first state that is wise enough to do it. We already know that the animals will live here. If we bring in clean, healthy, disease-free animals, there is no worry from disease; we know that, should any animal reproduce too much (which is unlikely with big game), we can send in an army of hunters to quickly kill them off; so what are we waiting for? The giraffe, in particular, would add a great deal to the exotic wildlife of our southern wilderness areas. It is one of the most spectacular of animals and the most colorful, making it a prime subject for photographers. In Kruger National Park in South Africa, only fifteen of these amazing animals were still living in 1902, because of incessant hunting. Now the animals have increased to over three thousand and have spilled out of the park into neighboring countries. The area in which they live is practically of the same longitude as Southern Florida, Southern Texas, and Southern California, except that it is the Southern Zone. But the animals will live in our country. This has already been proved over and over again.

I would like to see some states start an African big-game preserve right now—not on one or two thousand acres, but on a big piece of land of fifty to one hundred square miles. This way the land could be properly bounded and fenced without worry of the animals causing harm to humans or domestic livestock. There are many suitable areas for such a project in Florida and other states, and the land should be acquired now before it is subdivided into housing lots. In some areas it may be necessary to buy up a few dozen farms and houses, but these would not hurt a thing, for the buildings could be utilized for game ranger houses or for storing supplies, and the tilled land could be planted in grasses or crops that would be eaten by the animals.

Many people travel all the way to Africa to see animals in the wild. We could bring the very same herds to America where everyone could see them, not just the wealthy who can afford a trip abroad. I have been working on such a project for several years. Many experienced animal handlers and importers, such as Alton Freeman, Bill Chase, Richard Naegeli, Ivan T. Sanderson, V. Rider, Warren Prince, Harold Schultz, and Roland Lindemann, have all contributed freely of their time, answering numerous questions and offering the extreme value of their wide experience in the field. In addition, I have corresponded with zoo directors and wildlife park officials in Europe, the United States, and Africa. The accumulation of information that encompassed hundreds of long, detailed letters from all of the various sources, dozens of local and long-distance phone calls, and numerous trips to animal projects in Florida, Texas, California, New York, Michigan, and Africa have convinced me that such a project is absolutely feasible and practical.

I present here my own plan for a real jungle in America. It is presented for a piece of land of ten to fifty thousand acres. A smaller piece of land would not be suitable, for the intrusion of traffic noise would ruin the effect of being out in the jungles, away from civilization. The project could be built in Florida, Texas, or California, or perhaps Arizona or New Mexico. A much larger and more authentic version could be built in Central or South America, where vast areas of wild and undeveloped land are still available. A large island, a jungle valley, or perhaps a triangular piece of jungle land, separated and defined by rivers, would be suitable for such a project so that the African animals could not escape into the large jungles. In such an area as an Amazon Basin site could afford, even huge herds of hippos and elephants could be released. There are vast open grass areas around the jungles that, in some instances, are used as cattle ranches despite their remoteness. If cattle can survive there, then certainly elephants (especially the forest elephants), hippos, and rhino would live. There would be problems with disease, vampire bats, and the like at present, but everything has problems and they would need to be solved. But the tourist potential

A cape buffalo as viewed from Treetops. The author spent the night at the famous lodge, high up in the trees, and watched elephants, rhino, and buffalos confront each other at the salt lick directly below. A bright artificial "moon" lit up the area about the lodge so that the animals could be seen after dark when most of their activity takes place.

for such a project would be fantastic. In Africa, some of the smaller countries take in fifty million dollars a year on tourism, and all of it is based on people coming there to see the animals in the wild. People do not go to Africa to see the cities—we have cities in America. They go there to see the animals. A Central or South American jungle project could be a real tourist boon to the country that initiates it. But it must be done right, with authenticity. Treetop lodges could be constructed in the remote areas where guests could stay overnight to watch the animals congregate around a salt lick. If the project were built in the Amazon, the jungle could be preserved in its pristine condition, and treetop lodges could be constructed here so that visitors could stay overnight in the heart of the Amazon to see the unique South American fauna in its native jungle. The author feels that such lodges built high in the trees would eventually become as famous as the Treetops of Africa and that the nocturnal wildlife of the jungles could be lured from the bush by salt, fruit, or meat. Eventually ocelots, jaguars, and all of the wonderful animals and reptiles of the Amazon would make their appearance as they do in Africa's jungle lodges. In one jungle lodge in Africa, the Secret Valley Game Lodge, near Mt. Kenya, the elusive leopard has been lured from the jungle so that guests at the lodge can nearly always see them there. This is one of the few places in the world where you can practically be guaranteed to see leopards in the wild.

The Serengeti Plains in Africa is the home of hundreds of thousands of animals. For a few million dollars, we could build an exact duplicate of Serengeti in America, complete with acacia trees and all of the wonderful animals of the veldt. This could be built on public land, for the project would be for the public to enjoy.

A lion at Serengeti Park in Africa. Such a splendid scene could easily be duplicated in America. It would even be possible to keep lions in with antelope and other game as a natural predator, if the area were large enough and if the hoofed stock increased enough so that it would require natural cropping.

The nocturnal jaguar of South America should react the same way, so a treetop forest lodge in the Amazon jungles could present this powerful beast, as well as the ocelot and smaller jungle cats, as a main attraction. There are all kinds of interesting and unusual creatures in the Amazon jungles, and few are seen by the occasional traveler who wanders up the river by boat. But stay in one place, high up in the multistory level of gigantic trees in a comfortable treehouse, and you would see an enormous variety of wildlife, particularly at night when a dull light would illuminate the forest area around the lodge, as is done in the African lodges. The area for such a tree lodge would need to be carefully selected, perhaps overlooking a position of a river. In the high forests, as on the ground, animals have paths and trails, like *highways,* as noted naturalist Ivan T. Sanderson refers to them in his *Book of Great Jungles* (which should be read by anyone interested in the jungles). Also some areas are richer in wildlife than others. A treetop lodge in a true South American rain forest could become a fabulous tourist attraction, if the area around it were kept totally wild. There would be room for dozens of such lodges in the vast jungles of South and Central America.

In our own United States, a good duplication of the jungles could be constructed. The project would definitely work, for already similar exhibits on a smaller scale have proven this. The animals would thrive and even reproduce. As to the financial success of the project, this would be determined only by proper management. There is room for a really large jungle project in all of the southernmost states. Such a project would not harm the smaller exhibits, for people today usually go to the exhibit closest to them or to one that is directly on the road that they are traveling. It would take a millionaire or several people with a sustained income to do the project right. Perhaps the government should do it also. If there are more than one jungle project, each could be different so that people would go to all of them. A really big jungle project could give steady employment to thousands of people. There would be need for wardens, minibus or Landrover drivers, farmers, animal trappers, handlers, and even dancers, who would dress in authentic African costume and perform tribal dances for the entertainment of the guests.

A large jungle project would preserve a vast tract of land and save it from the developer. It would be a great thing for America. A number of people worked with me, attempting to get the project off the ground, but to this date, is is still a dream. Carl Selph, Russ Smiley, and Henry Deens all were interested in the project. It was Mr. Smiley who pointed out some of the choicest land in Central Florida for the preserve. Mr. Deens worked for a considerable time trying to raise the funds for the project to no avail. Many people were very interested in the potential of the project. In fact the mere mention of a real jungle in Florida or Texas would immediately draw stories from the press. But to raise the millions of dollars necessary to actually do the project right, to date, has been unsuccessful.

What follows is my original plan.

The Jungles

"See twenty-five wild elephants roaming free in a tropical forest. Huge herds of gnu, impalas, zebras, and gemsbok strolling across the plains. Photograph lions from your own car, or walk through the mysterious African rain forest and watch for monkeys, birds, and other wonders of the Dark Continent. Go for an elephant ride through the tall jungle grass or sleep in a treehouse deep in the jungle, overlooking a fantastic water hole, where the big game come to drink at the end of the day. This is The Jungles, America's greatest attraction."

There are millions of Americans who will never in their lifetime have the opportunity to visit Africa or see first hand the magnificent panorama of wildlife for which that continent is famous. The roar of a lion, the charge of a bull elephant, or the majestic giraffe may only be heard or seen in movies or in the close confines of steel and bars at the local zoo.

Only the wealthy can afford to make the journey to the Dark Continent, and even with wealth, the trip is not an easy one. Multitudes of red tape, customs, shots, passports—all time consuming—cut deeply into the trip. Africa is such a long way from our own country that even with modern jet travel, the trip is both extremely expensive and tiring. Why not bring Africa to the United States? In this day of atomic miracles, moon rockets, and supersonic planes, transplanting the fauna and flora of one continent to another is relatively simple and actually presents no major problems.

Already, many of Africa's most interesting plants have been successfully grown and propagated in our own country. The jungle vines, the pampas grass, bamboo, and many exotic trees grow and flourish in our country. The animals too will thrive outdoors, in the warmer southern states. In a similar enterprise, zebras and ostriches multiplied in large numbers when kept in a natural area in Florida.

It is a known fact that Africa's big game is rapidly disappearing, and the day may come in our lifetime that many of the large and spectacular beasts will become extremely rare or extinct on that continent. It is time that we bring breeding stock to the new world and propagate them under natural conditions to assure preservation of the species.

The purpose of The Jungles in Florida is twofold. One, it is to be a duplication of the jungles in Africa, complete with splashing hippos, chattering monkeys, thundering herds of gnu, eland, zebras, and impalas. It will have immense rain

Artist's concept of African Jungle, as proposed by the author. There would be an open veldt and dense hammocks, each with its own unique flora and fauna. A series of ponds, large and small, would be connected by a scenic waterway so that the animals could be viewed either by boat or Landrover. Deep in the forest area would be a treehouse hotel for overnight guests.

forests, dense swamps, and the veldt so familiar to the African tourist. It will give the American people an opportunity to see, hear, and photograph the spectacular fauna of a foreign land at relatively little expense. In addition, this fantastic undertaking will serve as a Noah's ark in the new land. The animals will reproduce their kind under the care and watchful eye of trained naturalists. Surplus species will replenish dwindling stock in the nation's zoos, for as our country grows, more zoos are needed, and of course a zoo cannot operate without animals. The Jungles will stir the imagination of young and old alike. It will be America's greatest attraction and fulfill the dreams of those who love nature and our wild animal friends. It will be a major conservation effort.

Modes of Travel through the Jungles

Travel through the jungle will be by Landrover, elephant, safari, or riverboat. Depending upon the land mass accumulated, private automobiles may be allowed in some areas. Little or no foot travel will be permitted except in nearby rain forests or jungle gardens containing no dangerous animals.

The Landrovers will be employed for photo-sightseeing trips across the veldt. Lions will roam free and the machines can approach them without worry or fear. The Landrovers will be used in all areas except where rhino, elephants, or water buffalo are living, and due to the unpredictable nature of these beasts, they will be viewed from overlooking embankments, which conceal escape-proof fences or cliffs.

Riverboats will ply the unexplored reaches of the Cameroons where giant hippos will cavort and huge crocodiles will line the banks. Flocks of pink flamingos will rise from the banks, and along the shore will be seen a real native village. The boat will dock here and the guests will photograph the natives. Some of the natives will be making spears. Others will be dancing or playing drums

Main building of African Jungle. The main building would be headquarters for all trips through the jungles, whether by Landrover, boat, or minibus. It would contain a Safari Room, a waiting area with a fireplace, giant murals of Africa, elephant guns, and native handiwork. A crescendo of native drums would announce the arrival and departure of all tours.

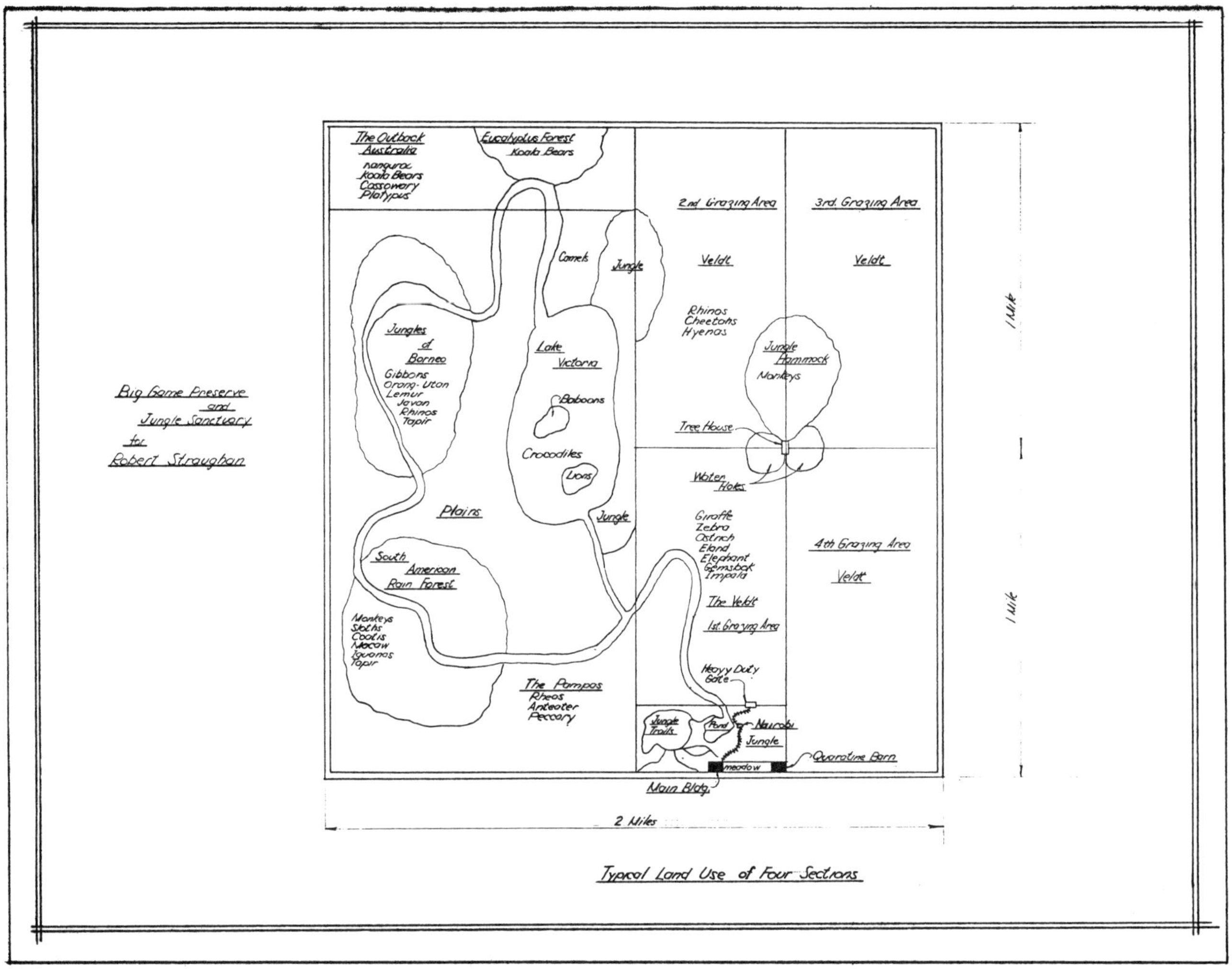

A typical land use of four sections of land, each one mile square. The two sections on the right are divided into four grazing compounds of hammock and jungle. Each area is one by one half mile, and all would be connected by gates so that animals could be driven from one area to the next to allow for reforestation and planting. The roads and tracks through the area would be far away from the fences so that the fences would never be seen. To the left would be one large fenced area with a winding waterway through various jungle areas. A double fence on the outside perimeter is shown.

or other instruments. After a brief visit, the boat will continue on through a vast swamp, filled with weird noises. Water buffalo and other large game will be visible on shore, and numerous islands will be visited, each with its own specific type of fauna. Baboons will inhabit one island, rhesus monkeys another. Various game animals that may not get along well together will be isolated on separate islands with hidden fences to discourage swimming from one island to another. The riverboats will be sturdy and covered completely with heavy fencing so that the people inside can never be harmed, should a group of monkeys try to climb aboard. This will also keep anyone from leaning too far overboard. The wire will be wide enough so that photos may be taken through it, or small

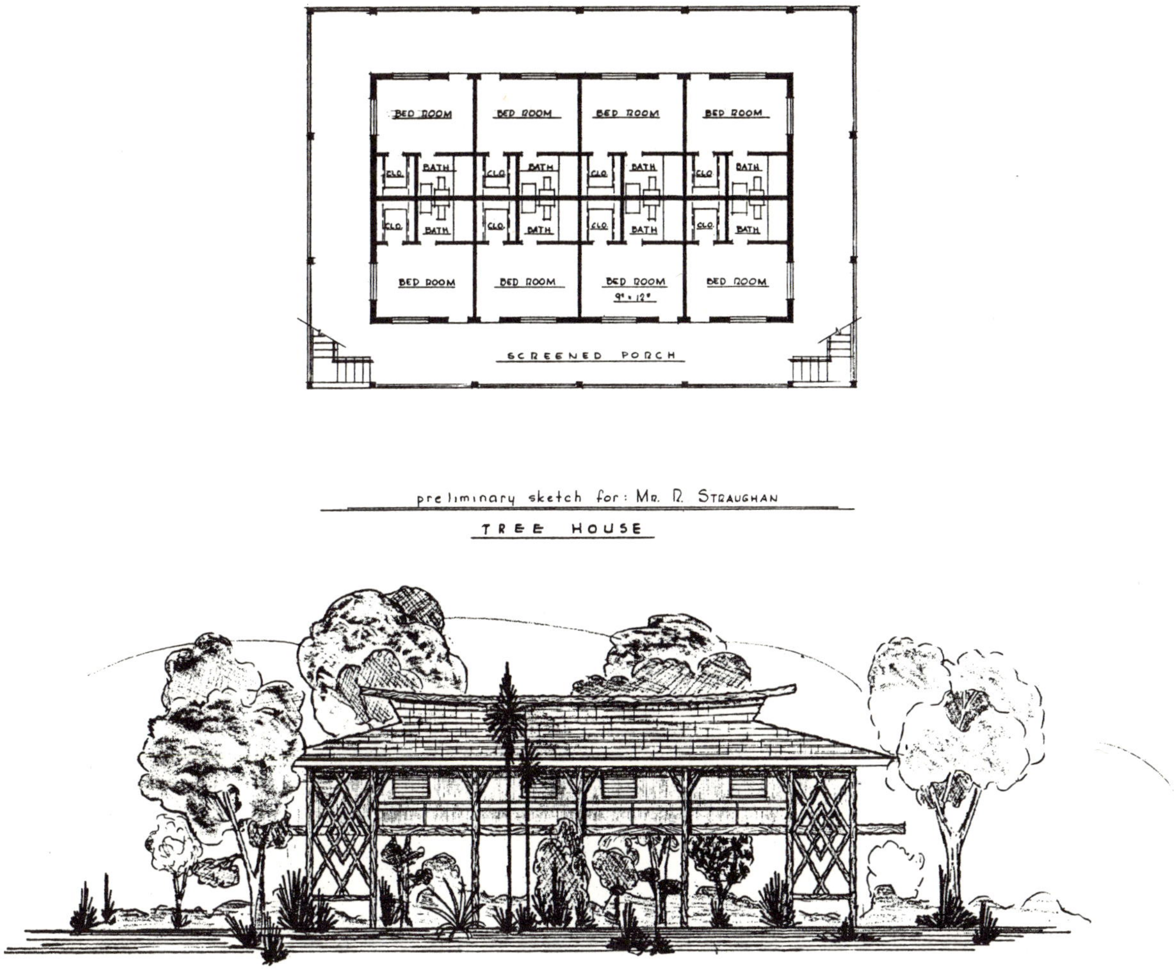

The treehouse lodge would be erected in a dense forest or jungle area among the branches of the trees in such a way that it would overlook a waterhole or salt lick area. It would have one or two stories of sleeping quarters with a walk-around screened porch. On the roof would be the dining area with a walk-around outside viewing area. The structure would be constructed on logs and pillars about thirty to forty feet above ground, depending on the area. It should not be anchored to the trees, but merely built among them, so that the wildlife may be viewed from all around.

slots or windows will be designed for this purpose.

The riverboats will have a huge area in which to operate. They will be designed after the old African trading boats, and the lengths of the trip will vary from a half hour to a full day, depending on the specific area for which they are to be used. The boats may be quite large, hauling twenty to thirty people at a time in comfort and safety.

All trips, whether by safari, riverboat, or Landrover, will be plotted and timed so that none ever comes in sight of the other unless such view enhances or highlights a trip. The magnitude of The Jungles must be such that all sounds of civilization are completely nonexistent. Distant auto horns, wailing ambulances, and other depressing sounds of the civilized world must be left behind. Even air travel over the area should be discouraged. Visitors must feel that they are really in the jungle, and everything possible will be done to achieve this effect. Ten thousand acres of land may be acquired for the project so that adequate breeding areas are assured.

Safety of the public will be stressed at all times, and guides will accompany all guests to insure their protection. Actually, a trip through The Jungles will be as safe or safer than a trip through the Smoky Mountains, bears included.

There will be so many different types of trips and exhibits at The Jungles that the tourist may spend as little or as much time as will fit his schedule. He may spend a half hour, a day, or a week. It will be like a trip to the jungles of darkest Africa, an unforgettable experience.

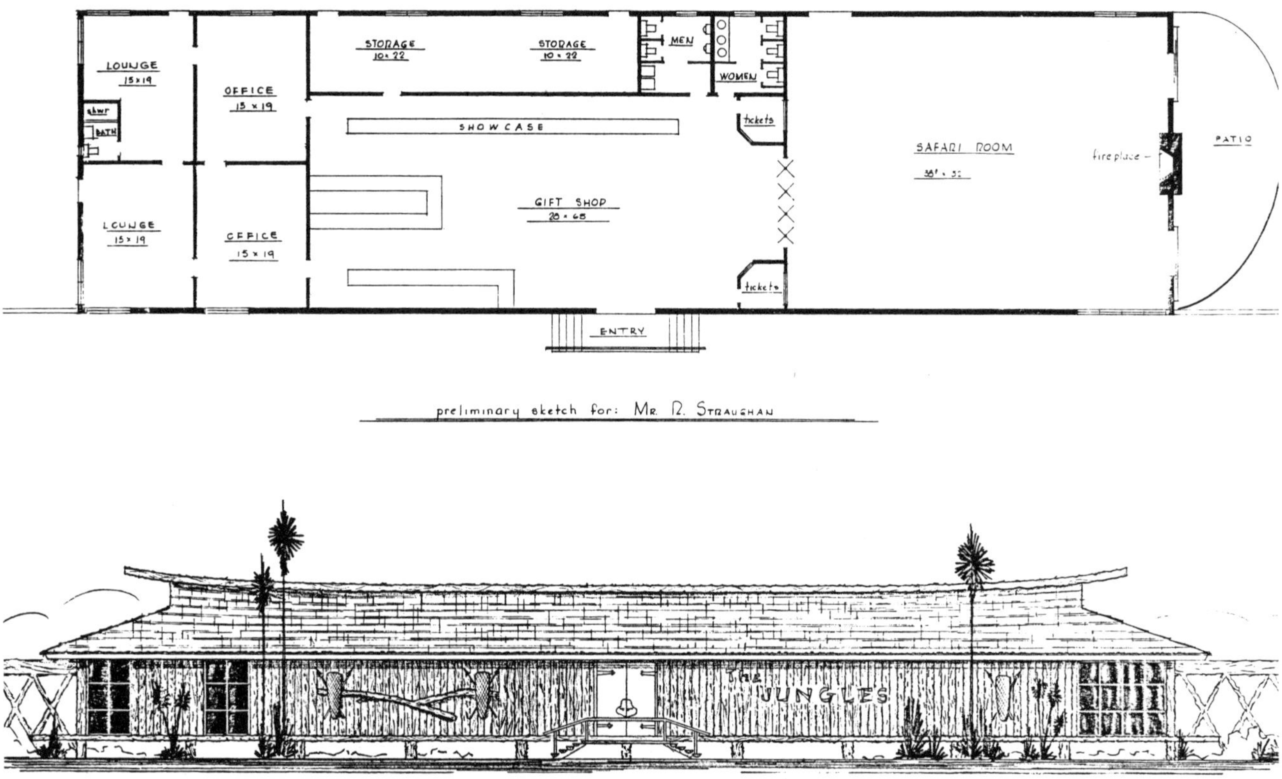

The main building of African Jungle would be headquarters for all cruises and tours. In a patio area outside the building, baby lions, small elephants, and the young of various species would roam freely for guests to photograph and enjoy.

Elephant Ride

An exciting method of travel through The Jungles will be on the back of elephants specially trained for this purpose. The elephants will carry small groups of tourists on carefully selected paths, through dense pampas grass six to eight feet tall, then through a wild, almost impenetrable rain forest, and continue on to the African veldt, where

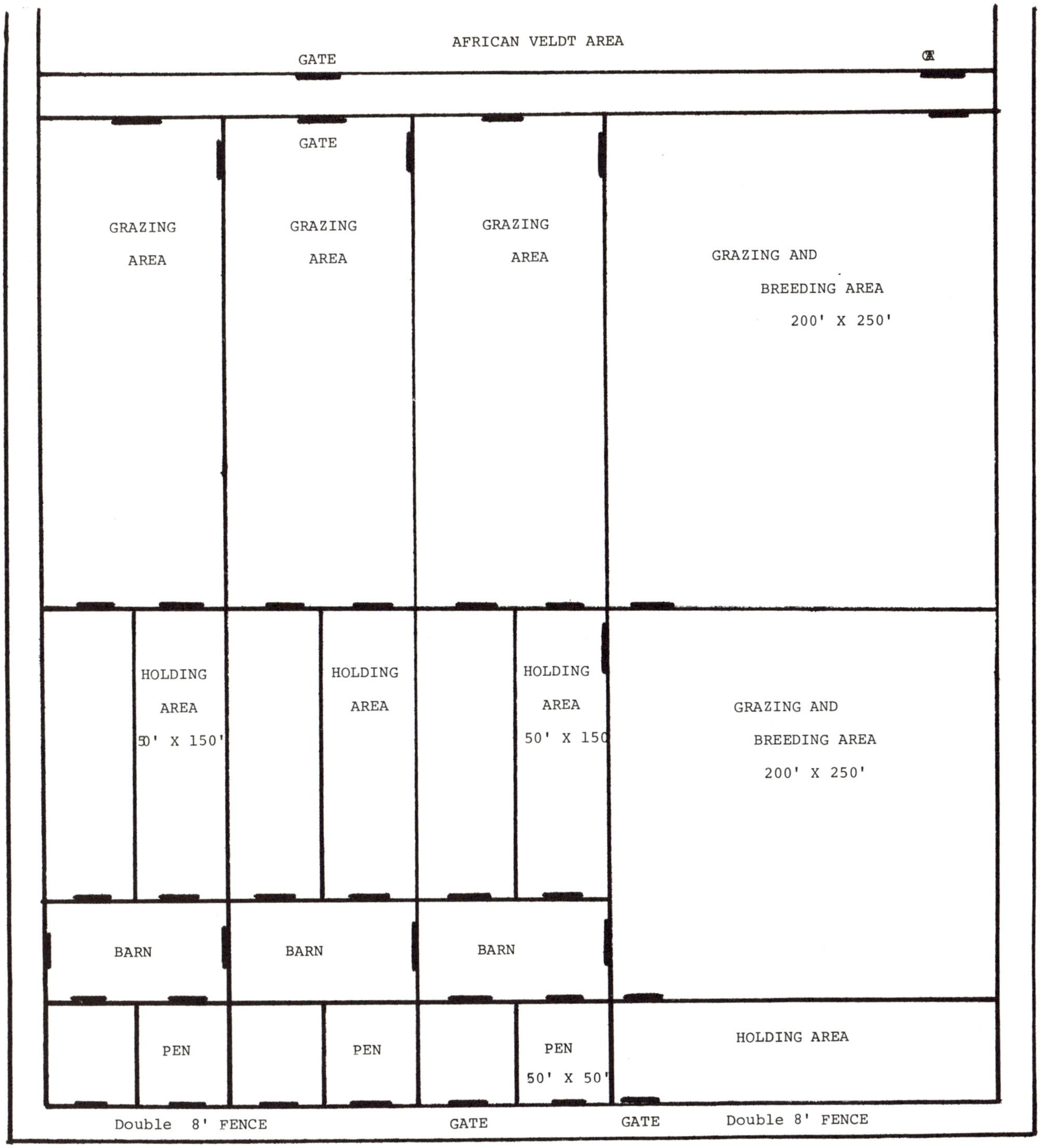

Introduction, quarantine, and breeding area for hoofed stock. This area would allow complete control and movement of the animals without actual handling of them. Animals are brought into barns or holding pens for inspection and treatment before finally being released into the vast veldt area. Gates between double fences will be wide enough to block passageways when opened, making a pass through the area for the animals. With this design, an animal can be moved from any pen or grazing area simply by opening the gates and gently driving the animal through. Special foam rubber padding may be used in some of the pens or holding areas where needed to protect overexcited animals from harm, should they run into the fence.

immense herds of zebras, gnu, gazelles, ostriches, and giraffes will be plainly visible on the open plains.

Guests will be able to photograph wild game from the height and safety of a huge elephant, giving them a most unique ride and a thrill beyond words. The *mahout,* or elephant driver, will also act as guide, naming the many species of animals that parade across the open plains as the spectacular panorama of wildlife unfolds before your very eyes. Deep in the jungle, your giant behemoth may pause while you photograph a band of baboons feeding on a wild fruit tree, or you may at last be able to see eye to eye with a huge giraffe, one of nature's most awesome creatures. Perhaps a photographer concealed in the bush will get a picture of you on your monstrous perch to make the occasion even more memorable. At any rate, an elephant ride through The Jungles will probably be the most exhilarating experience of your entire vacation. You will be safer on the back of this gentle beast than you would be in your own car and you will have no flat tires or traffic problems. In The Jungles you will be nearly as far removed from civilization as if you were on the moon.

The Safari

The safari will be an overnight journey to the farthest recess of the jungle area and will include a total, carefully planned itinerary of all the major attractions and points of interest in the jungle. It will be led by a white hunter in full regalia, and porters will carry supplies and equipment necessary for the trip.

A safari will consist of a dozen guests, more or less, who have the desire to spend a night in the heart of the jungle. The journey will be timed so that the party arrives at the water hole and treehouse just before dark. They will hear distant native drums, the roar of a lion, and authentic noises, both natural and recorded, as required to give complete authenticity to the atmosphere. Treetops, the famous jungle treehouse of Africa, will be duplicated here.

Erected in a huge banyan tree will be the treehouse, built high above the ground for protection from the prowling beasts. The house will be primitive but comfortable and will overlook the water hole, which will be the main center of attraction. Guests will be able to photograph the wild animals as they wander up to the water hole in small groups, as they do in their native land.

As nightfall approaches, guests may roam about the screened porch that goes completely around the house or sit in the game room, which will be handsomely furnished with authentic African handicraft. The jungle sounds will lull them to sleep or keep them wide awake, whichever is their cup of tea.

Safari Room and Gift Shop

At the entrance to The Jungles will be located the Safari Room, a luxuriously appointed waiting room where guests may sit and relax while waiting for friends who have gone on various tours. The room will be lavishly decorated with authentic jungle souvenirs and artifacts, many of which will be offered for sale in the gift shop.

The Safari Room will be headquarters for all of the trips throughout the project, and it will be a spacious, airy building constructed of huge, rough timber indicative of a jungle frontier structure. It will be a museum in itself with displays of native spears, shields, witch doctor masks, elephant guns, and many items associated with the jungles. Instead of the familiar popular music that bombards us everywhere, the faint sound of the tom-tom combined with actual recordings of jungle noises will set the atmosphere for adventures that lie ahead.

The gift shop will also be located near the entrance and will be in close proximity or adjoining the Safari Room. Only real, authentic arts and crafts from the jungles will be sold when practical. In addition, rare jungle plants, palms, ferns, and orchids will be on sale. Also, monkeys, birds, harmless reptiles, and small mammals will be offered to those who desire a live pet. Bamboo cages and other supplies in the pet line will also be sold. Film, both still and movie, will be a major sale item.

A spectacular tropical garden with giant ferns, orchids, bamboo forests, and huge elephant ears and other tropical plants will be attached to both the gift shop and the Safari Room. This area, which will encompass several acres, may be fenced with a ten- to eighteen-foot bamboo or wire fence and may be covered across the top with a heavy screen. Many colorful, exotic birds will fly free in the area, and the screen will be planted with vines so that it is not visible. Colorful marmoset monkeys, African chameleons, basilisk lizards, macaws, and many small, colorful birds will readily be seen by visitors walking through the well-defined paths. The purpose of this garden will be to display rare birds that ordinarily would fly away or be so shy that they could not be approached closely enough to see out in the wild. Also, it will give the people something to do while waiting for guided tours.

Spectacular undeveloped land in Florida that is suitable for a big-game preserve. The author feels that fifty square miles of this land should be purchased to build an authentic wild-animal preserve. It would be a tremendous conservation effort and would enable millions of Americans to see elephants, rhino, giraffes, and other animals living in the wild.

All points of departure and return will be just outside the Safari Room at a staging area known as Nairobi, the well-known African big-game center. Both riverboats and Landrovers will use this point as a junction for all trips, and the blast of an old steam whistle or the crescendo of a native drum will announce the impending departure of a trip. Thatched wooden structures and ancient signs throughout the compound will give the attraction an authentic, completely natural appearance. Employees, both men and women, will wear khaki uniforms, boots, and pith helmets, reminiscent of old Africa. Once inside the gates, a visitor will feel that he is on a real safari, and the worries and boredom of civilization will be left far behind.

In some areas of the preserve, people will be able to drive through in their own cars at a regulated 15 miles per hour much as they do in Kruger Park in South Africa.

Disease Control in the Jungle

Daily supervision and observation of all game animals will be made by a trained veterinarian, specially selected for work with wild animals. All animals will be duly quarantined and inspected for infection, ticks, and other signs of pests or disease, before release on the range. Animals already at large may be subdued and treated with hypo charges fired by attendants. This method is already in use both in the United States and in Africa by game wardens and scientists who wish to examine or brand wild creatures for study.

Spectacular African-veldtlike land in Southern Florida, in the area in which the author proposes to build his big-game preserve. Just picture giraffes and elephants emerging from the bush. The area would be divided so that the animals would do little damage to the foliage. Landrovers would drive people out into the bush to search for animals. All fences would be hidden from view. It would be just like a trip to Africa.

Animals of The Jungles will always be confined, and their enclosures will be made so that there is not even the remotest chance that they could ever escape to the surrounding land. First of all, the entire area will be completely surrounded by at least one or two deep canals that will be fenced as an added precaution, should any game escape its enclosure. The most dangerous animals will be well within the boundaries of the area so that should they escape one area, they would immediately be confined to the next. Trained animal handlers will always be on hand to manage such situations. Even large animals like elephants or rhino may be captured, if necessary, by a special double-gate trap that will connect the two sections of their compound. Since elephants and rhinoceros are known to uproot trees and plants, their compounds will be made into two equal sections so that when one area becomes unsightly, a gate will be opened, allowing them to move into the other section so that the first area may be cleaned and replanted as necessary. The area may also be sprayed with lice killer or tick killer if needed. The same system may be used with the plains animals. If one area becomes overgrazed or unsightly, the animals would be driven into a new area while the original area is replanted. A twelve-foot chain-link fence will encompass the entire property containing the animals. This will be in addition to inside fences and moats.

The Jungles as a Game Preserve

The Jungles will create tremendous interest for the general public by preserving many of the disappearing fauna of the world. Elephants, rhinoceros, giraffes, lions, water buffalo, zebras, hippopotamuses, and various antelope can be propagated under natural conditions in Florida. It is commonly agreed among animal dealers that such a project would be financially very profitable. There is an increasing demand for all of Africa's big game for the nation's zoos, not to mention South America and possibly Europe.

Given open range and protection against predators, all the exotic wild creatures of Africa could be raised under controlled conditions. Browsing animals in particular will flourish with sufficient water, shade, and supplementary foods supplied as needed. Special grasses, tropical fruit, hay, grains, and other foods can be grown right on the property by experienced farmers. In addition, certain areas of the open range can be closed off to game while it is plowed and planted in suitable grass. This will not only beautify an area but, more important, will ensure the animals of more than enough food. With proper management, game per acre can greatly exceed that of the wild state, making a harvest in zoo animals a profitable achievement.

Except for the lions and leopards, which are meat eaters, practically all the game animals will graze, which greatly simplifies their upkeep.

As a tourist attraction, The Jungles will be an immediate hit, for there is nothing like it in America. While the wild game is running free, increasing in number, millions of tourists will want to see it and photograph it. It is a known fact that

More of the junglelike land in Southern Florida. The author feels that all kinds of African animals would easily survive here, even through the cool winter. This beautiful area would provide shade, privacy, and protection from the weather. Authentic African baobabs and acacia would be placed in some areas, and wild fruit would be planted to provide natural food for the monkeys and other animals.

Africa's wildlife is rapidly diminishing. The Jungles will in itself become a major conservation effort, a most worthwhile cause.

Sources of Income for The Jungles

Financial requirements for the maintenance of The Jungles could be met by:

1. Charge for admission and for numerous trips of varying length and interest.
2. Use of The Jungles as a photographic location for a movie company or television interests.

3. Propagation and sale of animals, reptiles, and birds.
4. Sale of gifts, film, and pets, and operation of a restaurant.
5. Charge for a special safari to the treehouse overlooking the water hole.
6. Private guide service for photographers.
7. Motel units on the property.

I was so enthused about the prospect of a giant big-game preserve in our country, that I even wrote to the president about it, for financial aid. Here is the actual letter:

12-18-68

President Elect Richard Nixon
Washington, D.C.

Dear President Nixon,

In one of your election speeches you said that if we had any ideas that we felt would be good for the country, we should write to you about them.

I have been working on a great plan for many years which will capture the imagination of millions of people.

My idea is to create an African Big Game Preserve in Central Florida, Southern Texas and Southern California.

Millions of Americans will never in their lifetime have the opportunity to visit Africa to see elephant, giraffe, rhino and other big game animals roaming wild and free. They will never have a chance to go on safari in a land rover through the bush and marvel at the vast game herds for which that continent is famous.

So why not bring Africa and the jungles to the United States? For the cost of a single rocket now being fired into space, America could have actual herds of elephant, zebra, eland and hippopotamus. We could have an absolute, authentic jungle with even the acacia and baobab trees and typical veldt vegetation. Not a fence or wire would be seen so that the people going on safari would feel as though they were deep in the African bush.

The project would be a major conservation effort for we could assure the nation's zoos of a permanent supply of animals. In addition, it would offer safe refuge for certain animals now facing extinction in their native land.

The big game preserves should be built now, while there is still an opportunity to acquire large tracts of land in Florida as well as the other states. It should be done on a large tract of land, 100 to 300 square miles so that the animals will have sufficient grazing area. I have a complete prospective on it. If you are interested, I will send it.

Sincerely Yours,
Robert P. L. Straughan, Editor
Salt Water Aquarium Magazine

I received no reply. But perhaps someone else will think it is a pretty good idea. Sooner or later we will have that big-game preserve. Just wait and see!

Undeveloped hammock land in Southern Florida. The author wanted to build his big-game preserve on this land but could not raise the necessary funds. This land is ideal for such a project, for it has dense jungles, open veldt, water, and rich grass. There are about one hundred thousand acres of such land in this area.

First Steps in Building a Jungle or a Preserve

1. Selection of the Land

This of course is a major step, and the total acreage depends upon the size and scope of the project planned as well as on the specific fauna that it will contain. If you plan to raise and exhibit only lizards, you could get away with a small tract of ten to fifty acres. But a real jungle should have several thousand to ten thousand or more acres to create the proper atmosphere and provide sufficient grazing room for the animals. The land need not be close to any town. In fact if it is far off the beaten path, on a dirt road, it will be even more appealing to those who really want to see it. After all, you could hardly have a real jungle with cars passing by every few minutes on a densely traveled highway. If your jungle is authentic and well presented, people will beat a path to it. Cypress Gardens is a classic example of this. It was built completely off the main roads, yet is one of the most popular and successful attractions in America. Building a jungle in the "boondocks" has other worthwhile features. The land is certain to be cheaper there, there will be fewer problems with zoning and restrictions, and the quiet atmosphere will be better for the animals. And of course, the jungle has to be out in the wild.

It must be remembered that with any attraction, large or small, the tourist must be sold on it before he even gets to it, or he will drive past it, regardless of how well it is done. In other words, it must be a destination, planned in advance by the potential visitor. Even if you had live dinosaurs in your exhibit, most tourists would drive right past it, unless your project were on their schedule as a stopping place for their too brief vacation. They must read and hear about your project in magazines and newspapers and be sold on it before they leave home. This is why good press and public relations is so essential to the success of any attraction. It also points out the fact that you need not be directly on a highway, if your project is any good. A simple *Turn Here* sign on the highway will direct people to your project. If it is at the far end of a heavily wooded, dirt road, ten or fifteen miles from the main highway, by the time the people reach the main gates, they will be in the mood for a real jungle, having already passed through a wilderness area. This can add greatly to your project, particularly if you purchase the land on both sides to keep it wild, and have good signs every few miles assuring your guests that they are on the right road.

The farther south you can get for the project the better. Jungle hammock land with open fields is best, and at this writing, there are a fair amount of large, ten-thousand-acre tracts offered for sale in Florida, most of which would be suitable for a jungle project. Large trees are an invaluable asset, and your land should contain good areas of dense hammock. You can increase the jungle effect of these by planting exotic ficus trees, acacias, lianas, epiphytes, palms, and other tropical growths. In fact, if you locate in Southern Florida, you could build an immense, almost impenetrable jungle by planting a few hundred medium-sized ficus trees (*Ficus bengalensis*), which is the most spectacular of all trees, twenty or thirty feet apart. They would soon grow together into a mass of twisted limbs and giant branches. They are a botanical wonder. Other ficus trees are equally spectacular and grow to enormous size. You could obtain small to medium trees in quantity for a relatively small sum, and with proper planting, watering, and fertilizing, they would grow quite large in five years or so.

Your land also should contain small ponds, lakes, or spring-fed streams, since water is essential, not only for drinking and irrigation of the grasses and trees but also for fire protection. One serious fire could destroy your project, killing all the animals, and would be a major catastrophe. Land that is surrounded by water, particularly hardwood forests surrounded by streams, would practically eliminate any serious threat of fire. The land should be purchased for a big-game preserve, and this should be in the deed or legal contract as protection from zoning regulations. If possible, do not locate near cattle country or horse farms, because the owners are likely to object to any hoofed animals in your project, for fear of ticks or disease. After you have acquired the land, the next step will be fencing.

2. Fencing the Land

First, take low-altitude aerial photos of the land so that you can plot out the land to its best advantage, marking the areas to be divided, ponds, connecting channels, jungle and veldt area. Then you will be ready for your outside perimeter fences. You will need double fences, one set twenty feet apart from the other. These should be ten- to twelve-feet high and buried three feet into the ground, preferably anchored in cement. Fence specialist Fred Leonard suggests that the area of the fence above ground, where it meets the wire, should be coated with plastic spray, because the cement may eat away the galvanizing. Also, the cement should come to a point at the fence, about one foot above ground. This will keep water from settling there and eventually rusting the wire.

If finances permit it, the outer fences should be constructed as soon as you have photographed and surveyed the land both by air and by Land-rover. There should be an additional fence five hundred feet in from the double fences, at the main entrance, as a safety fence so that animals will not escape when the main gates are opened and closed, especially at the start of the project when a great deal of machinery will be passing back and forth.

As soon as the double fences are completed, you can start releasing animals into the area if they are zoo-bred or locally raised animals, which would be free of disease and ticks. All kinds of

Beautiful, wild, African-veldtlike land still exists in South-Central Florida. Before it is developed into housing lots, huge tracts of it should be bought up and made into animal preserves.

animals, from giraffes to zebras, even baby elephants, could be turned loose. Then, as soon as you have a sufficient number and variety of them, you could open your gates for admissions, even while you are building the project. This would bring in income from the very start. You could mark off roads or tracts and let people drive through the area in their own cars, observing a strict fifteen-mile-per-hour speed limit, or depending on the area you might find it profitable to take guests through in a Landrover or a minibus, as is done in Africa.

3. Administration Building

This would be your next step. You could start your main building while the fence is going up. The main building would contain offices, rest rooms, a gift shop, a ticket office, and safari waiting room. It need not be an expensive building. In fact, I had planned one of logs, which would be impressive but quite inexpensive. If it looks a little rugged and unfinished, it could add to the jungle theme.

4. Animal Quarantine Area

This should be built as soon as possible and would consist of a number of barns, stalls, feed storage sheds, and numerous paddocks so that animals could be confined, inspected, and treated for disease before they are released into the large area. It would also contain a lab, egg-hatching area, and outdoor paddocks, both large and small, so that animals could be held there for weeks or months if necessary. No animal or bird should be released into the large outdoor area until it has been inspected by a qualified animal expert. One sick animal can raise havoc with all of your livestock, so extreme caution must be used when acquiring and releasing animals. Imported animals in particular must be inspected with great care. Native-born zoo animals will most likely be safe, since they will have been scrutinized for disease over a long period, but even these should be checked by your own animal handlers before release.

5. Additional Fences and Partitions

The area-dividing fences should be constructed as soon as possible so that you will have some control over the animals. You will have to decide whether to divide your land into halves, fourths, or eighths, but you will most likely need some division of your land so that you can drive the hoofed stock from one area to another, for replanting or spraying of an area. If an area becomes unsightly from overgrazing, tire tracks, or denuded vegetation, you might wish to replant it, or you may need to plow up certain fields to plant in tall grasses as food for the many animals. This would not interfere with the operation of your project, for you would simply close off the area to tourists and take them through the new area. Trees and shrubs can be a valuable asset to a jungle, and care must be taken that they are not destroyed by the animals. If you plant fifty or one hundred acres of sugarcane or tall sudan grass, your elephants and other grazers will (hopefully) feed here, but if you have no natural food for them, they will quickly pull down the branches of even your tallest trees. The gates from one area to the next can be constructed of heavy beams, and the fencing near the gate could be planted in dense bamboo so that the fence would not even be seen. Passing through an enormous gate of logs would add suspense as to the giant animals on the other side that necessitate such a strong structure.

6. Animal Trapping Areas

It will be necessary to capture and subdue various animals from time to time, either to medicate them or to capture them for shipment. You can make a number of natural trapping areas at the corners of your fences. This is done by eliminating the corners, rounding them off with extra wire so that the animals will not hover in the corner should they follow the fence. The inside wire used to round off the fence is not anchored to the ground but simply fastened to pipes that fit into slots in the ground. This way, the fence can be removed at one end and swung out from the main fence when animals are to be trapped. The animals are simply driven into the triangular pocket between the inside wire and the main fence. Boxes or crates are set into holes in the main fence and the animals are easily captured, after which the removable wire fence is set back in place. Or if your fencing has no corners, you can utilize the same effect by erecting a temporary fence at a right angle to the main fence with open areas through which the animals can be driven. A good cowhand or rancher can assist you here with numerous ideas.

The popular dart gun may be used under certain conditions, but it is not infallible. In fact it sometimes kills more animals than it saves. Rangers working with the rare Florida Keys deer stated that they stopped using it altogether because it caused the death of five of the endangered animals. I wonder how many animals have been killed in African game preserves and never reported by researchers, who seem to find it necessary to subdue and tag every animal in Africa in their studies. I think a little darting is necessary but that better study of animals can be done with binoculars. I sometimes feel that the dart gun kills more animals than it saves. There are other ways to capture animals.

It might be feasible to construct lion or tiger dens from large concrete pipe, six or eight feet in diameter. A T-shaped joint could be fitted with barred gates that could be opened and closed from the safety of a Landrover. The pipe could be covered with large boulders or soil, leaving only the dark opening exposed. The animals could be fed their choice meals there, which would encourage or train them to visit the area. This would allow easy trapping of them or locking them up at night, should this be desired. Animals that are to be collected could be driven from the enclosure through an adjoining or selecting cage. Large rocks placed upon such a structure could be painted flat black so that in winter they would absorb the heat from the sun and tend to keep the area below warmer during cold nights. This would also help keep the area warm during brief cold spells, for large, dark-colored rocks will retain heat for a long while. Dark-colored rocks could also be used as warm dens for many of the smaller animals, as well as for lizards and turtles.

Many small mammals, monkeys, and birds will nest at night in a hollow tree, so by providing them with a real-looking hollow log suspended from a tree branch or set in the fork of a tree, you will be giving them a home. You can also capture

A wild grass and hammock area in Central Florida that would be ideal for an American big-game preserve. The area has everything: dense woods, open plains, scrub brush, and swamp. The author would like to see a preserve built here only if the area is kept wild and untamed with no signs of civilization except for game-viewing lodges hidden deep in the bush.

them by attaching an easily tripped lid to the hollowed log. At night, you can visit the area and close the lid on the log houses of the specimens that you wish to collect. Then bring the house and all back to the examination or shipping area.

7. Publicity and Promotion

Your jungle project is not likely to be a financial success without proper advertising, promotion, and management. Newspaper and magazine articles are very helpful, and radio or television should also be used, particularly if you are trying to interest local people. Billboards are nice also, but national magazine coverage, particularly sporting, animal, and conservation magazines, will help you a great deal. In fact, you could publish a small monthly news bulletin with photos and news of happenings in your jungle. This could be a six-by-nine-inch magazine with eight pages, and you could send it to all magazines and news media as a constant reminder of your most worthwhile project. It would cost little to produce and would be worth its weight in gold. I have published my own thirty-two-page magazine for many years, and it has never been a big chore, despite all of my other activities. It has been well worth the effort.

8. Major Equipment Needed

A ditch digger, a fence erection crew, bulldozers, and draglines for making the ponds and channels will all be needed. Most of it would simply be rented for the jobs. But several large irrigation and water-pumping machines may be needed permanently to water the hammock areas and perhaps move water through the many streams and channels. You may wish to create a waterfall, a miniature Mt. Kilimanjaro, or an Ngorongoro Crater. Fire fighting equipment should not be neglected and the amount needed will depend upon your location.

9. Ponds and Channels

You may find an area already dotted with

Like an artist's painting, a cypress swamp is a thing of exquisite beauty. Many such areas appear in the land selected by the author for his proposed big-game preserve. Guests would be poled through the area in boats, or electric motors would be used on the boats to prevent noise and water pollution. Many areas would be kept inviolate except for added alligators, crocodiles, and other animals that would fit in with the area's ecology. The natural beauty would always be preserved.

natural ponds or rivers. This will be fine and will add much to the beauty of your jungle. In Central Florida, many large tracts of land contain numerous lakes and spring-fed streams. These are excellent for cruises to various parts of your jungle. But if there are few or no ponds on your land, you can dig good-sized ponds and even lakes and connect them with a deep, wide, winding, natural looking canal or river. If you do construct ponds and lakes, you could leave large islands in the center of them that later could be stocked with gorillas, chimps, gibbons, and other valuable animals. These animals cannot swim, so they would stay on the islands. You could plant a number of fruit trees on the islands to furnish them with natural food. The islands should be of large, natural hammocks with huge trees and should encompass ten to fifty acres each so that the animals will have sufficient room to forage. The animals can easily be enticed out into the open by bringing food to them.

In low areas, such as much of Southern Florida, it would be wise to use the fill from the ponds and lakes to build high mounds and hills near the ponds and canals. These can be sodded over with grass and eventually planted with trees. They will give the animals protection from occasional floods that might occur and in addition they will add scenic beauty to the flat areas. They should tower at least thirty feet above the surrounding area or even higher, since hurricanes sometimes raise the water quite high. If you plan a Lake Victoria on your project and must dig it, you will have enough fill from the hole to build a small Mt. Kilimanjaro. This could be a few hundred feet high and the top could be covered with fine, white sand, which would look like snow from the distance. Thus you would be utilizing the soil and gravel taken from the lake bottom and would be creating two attractions at the same time for your project. A large lake with a huge herd of hippos would certainly be of great interest for any jungle project. Make certain to plant the surrounding areas with tall sugarcane and grasses to provide food for them. They feed mostly on land and rarely eat water plants so that you must provide good grazing for them.

The ponds, rivers, or canals will provide a means with which to view the animals in comfort and safety. Large *African Queen*-type boats could carry tourists through much of the area and provide a safe yet scenic cruise through even the densest jungle. In addition, they are a real fire-safety factor. They also can be used to isolate nonswimming animals, confining them to specific areas.

10. Treehouse and Game Lodges

If your area is large enough, you may wish to construct one or several treetop lodges and game lodges in remote areas of the jungles. The treehouses should be built on pilings or logs, among but not attached to the trees. Since trees grow unevenly, and also since they sway considerably, it is best to build the structures separately from them. Overnight guests could stay in the treehouse or remote game lodge where they could observe and study the animals at night, with the aid of a light that would illuminate the immediate area. A salt lick nearby would lure the animals to the area. This is a fascinating way to observe animals, and the Treetops Hotel in Africa is so popular that it is solidly booked all of the time.

The game lodges would be for daylight as well as nighttime viewing and would be constructed on a hill overlooking a large veldt area. Guests would stay there for several days at a time, and the lodge would be located in a remote section of

A beautiful, natural alligator exhibit, called Gatorama, in Southern Florida. Hundreds of the reptiles swim in a large pond, snapping at anything that falls into the water. This is a much more effective way to keep and display alligators than the usual smelly pit. The reptiles live as naturally as if they were in the wild. This scene could be duplicated in a ten-thousand-acre jungle. Guests could view the reptiles by boat.

the preserve that would not be visited by daily traffic. In the evening, a roaring campfire would add to the beauty of a night in the jungle. Both the game lodge and treetop houses could be constructed while the rest of the project was getting under way.

It must be pointed out that it may take a year or several years to accumulate a really effective stock of animals, so it will be necessary to work on various portions of the project while awaiting the animals. There are many complications to building a big-game preserve like the one that I envision, and I do not claim to have all of the answers. I am merely presenting a number of ideas and suggestions for such a project, basing most of them on operations and projects that have already been successful.

To be really effective, you will need *herds* of African animals. A few hundred animals on ten thousand acres of land would be like a drop in the bucket, and you would have to search all day simply to find them. But the careful and wise selection of animals at the start will be very important, particularly if you intend to open the gates for admission. Spectacular animals such as giraffes, zebras, and elephants can easily be seen, even at a distance. These three animals alone, if supplied in number, will give a good start to an African jungle. Add forty or fifty ostriches, some camels, a few dozen elands (if you can get them), and as many large antelope as you can obtain, as well as white rhino, and you should have enough to make admission to the project feasible. The fact that you cannot see all of the animals at once and must look for them has a certain amount of appeal in itself. This is the major appeal of Kruger Park in Africa, where the animals are all hidden by dense underbrush. You must look for the animals as you drive along deserted roads. The excitement of finding a lion or a kudu hidden in the bush is quite a thrill. If it were sitting out in plain view in an open field, seeing it would lose much of its impact. Thus, a few gorillas in a dense jungle area on an island would present a thrill to people who would need to search actively for them. If you make the island too small, it will be worthless and even look a little foolish as you approach it with a boat. A large island with hills, large rocks, and dense trees gives the feeling of mystery. In addition it provides privacy for the animals, which they need if they are to survive and reproduce.

Of utmost importance is the hiding of the fences, if you are to portray the idea of a real jungle. Nothing deflates the impression quicker than seeing fences all about you. This is why the author suggests such large areas for the project. You just cannot really present an authentic jungle effect if you hear cars passing by, hear horns blowing, or see fences all around you. If a large area is divided into several sections of five hundred or one thousand acres each, or even five thousand acres each, twisting roads, trails, or canals would simply be out of view of the fences. Woods, hills, and high ridges of fill left over from digging of canals could hide fences from view. Dense growths of bamboo could hide fences near gates after one has entered the preserve. In Livingstone Game Park in Africa, I drove through wooded trails for the better part of an afternoon seeing giraffes, impalas, rhino, baboons, zebras, buffalo, and all kinds of game without once seeing a fence, except at the entrance, and yet the whole game park comprises only two square miles. All of the fences were simply hidden from view by the dense woods, giving the effect that we were in a vast wilderness. This one game preserve in particular has convinced me that my long-planned jungle project is entirely feasible, for all that was done to create it was simply to fence off an area of jungle-type land and stock it with animals. This is precisely what a jungle project could be like in America.

America could easily have a Kruger National Park, a Serengeti Plains, or an Ngorongoro Crater, filled with thundering herds of animals. These and more are all within the realm of possibility. All it takes is imagination, bold planning, money and the determined will to do it. Compared to going to the moon, it is a cinch.

12
Animals for a Jungle Preserve

The best place to obtain animals for a jungle exhibit or an open-air zoo is from other exhibits and zoos. Nearly all of them have an annual surplus list of animals and birds that they wish to exchange or sell. These animals are native born and are usually much hardier than imported stock. Also there are fewer complications in keeping them, from a legal standpoint. It should be remembered that only zoo-bred or native-born African animals of any hoofed stock such as antelope can be owned by anyone other than zoos. There are hundreds of zoos and animal farms and exhibits throughout the country, and it is possible to obtain a good variety of animals directly from them.

If they do not have what you want, then you would have to import the animal, but rather than do it yourself, you would be wise to work with established animal importers. They already have the contacts for the animals, and also, they are aware of all laws and restrictions pertaining to all animals. This can save you a great deal of time and disappointment. Also, it should be remembered that an animal picked up from a local importer at his place of business here in the United States is a surer thing than an animal offered at a much lower price in some foreign country. The animal offered to you in the jungles may be in excellent shape at its compound, but by the time it reaches you, it could be injured or very sick.

Most importers will work with you on a percentage basis if you will need a great many animals. Thus you will be able to take advantage of their experience, which is worth much more than the percentage they charge. There is much skill required in transporting animals from the jungle or veldt to the States, and there is nothing like experience. Experienced animal handlers are available in the trade, but you must seek them out because they are not overly abundant.

Financing of a jungle project is best done with a stock arrangement, for a sizable project can easily run into one million dollars or more. The huge projects that I envision could probably only be undertaken by a millionaire or a billionaire. Of course, the government could do it if it would utilize some of its land for that purpose. The major expense outside of the land is the fencing. This can be as expensive as the land itself, but it need not be purchased all at one time. Sections of an area could be fenced and these could be stocked as the animals become available. The owner of the Catskill Game Farm in New York advised me to start small with my project and let the animals reproduce to build up the stock. This is sound thinking from a place that is eminently successful. One could start out on one thousand acres with leases on additional land and begin with a nucleus of animals selected in pairs or threes (two females and a male).

I have compiled lists of animals and birds for various jungle projects with approximate prices of the specimens. Prices and availability of animals vary from year to year, and the sources from which you obtain them also vary. Sometimes a zoo or attraction will practically give you an elephant, hippo, or lion because they cannot keep it and will be happy to find a home for it. Also some zoos and animal farms will let you board their animals for them, free, if you will feed them, for the winter months especially if your project is located in a southern climate. They may also let you have surplus animals with the understanding that you give them some of the animals' young in future years when they need them.

Live Specimens for an African Jungle Exhibit

Select 150 to 1500 acres of veldt, open pasture with scrub brush, dense hammocks, and small ponds. A Southern climate is preferred. This would be just a starter. You would need three times as much to make a good showing.

African Stock

Specimen	Cost per specimen	Total cost
3 Rhinoceros, white	$6,000	$18,000
10 Ostriches	200	2,000
10 Gemsboks	1,000	10,000
3 Hippos	2,000	6,000
3 Giraffes	3,000	9,000
2 Elephants	3,000	6,000
8 Zebras	1,200	9,600
3 Aard wolves	100	300
5 Cheetahs	800	4,000
5 Lions	200	1,000
3 Kudus	1,500	4,500
5 Elands	600	3,000
2 Gnu	1,200	2,400
3 Camels	1,000	3,000
4 Hornbills	100	400
5 Secretary birds	200	1,000
10 Guinea fowl, exotic	80	800
10 Diana monkeys	100	1,000
10 Colobus monkeys	100	1,000
10 Mandrills	100	1,000
10 Chimpanzees	400	4,000
10 Baboons	100	1,000
10 Antelope, African	1,000	10,000
20 Chameleons, African	5	100
5 Crocodiles	40	200
6 Frogs, giant	30	180
175		$99,480

South American Rain Forest

Select 300 to 900 acres of heavy, dense, semi-tropical hammock land that would be surrounded by a wide, water moat and interspersed with boat channels. Daily spraying of the trees would increase the growth. Water from deep wells could keep the area warmer in winter. This could be a fabulous exhibit for Florida, Southern Texas, or California. Florida has many areas suitable for such an exhibit.

South American Stock

Specimen	Cost per specimen	Total cost
10 Giant anteaters	$100	$ 1,000
5 Tapirs	100	500
25 Macaws	40	1,000
50 Toucans	20	1,000
10 Cocks of the rock	50	500
50 Exotic birds	10	500
10 Sloth	25	250
5 Freshwater porpoises	400	2,000
10 Capybaras	20	200
5 Collared peccaries	40	200
10 Wooly monkeys	40	400
10 Capuchins	25	250
10 Howler monkeys	20	200
10 Spider monkeys	25	250
100 Squirrel monkeys	5	500
100 Marmosets & monkeys	8	800
6 Golden lion marmosets	35	210
100 Green iguanas	3	300
25 Caimans, medium size	3	75
10 Flamingos	80	800
561		$10,935

This would be the very minimum of stock and it would need to be conspicuously displayed until the stock is increased. Three times the amount would make a better showing.

Asian Jungles

Choose 300 to 900 acres of open pasture, dense hammocks with numerous ponds and woodlands in a semi-tropical area if possible. A natural waterway for boats would be an asset. A South Florida location would be best. The Malay tapir should also be included if obtainable.

Indo-Pacific Stock (Borneo, India, Etc.)

Specimen	Cost per specimen	Total cost
3 Giant tortoises	$400	$ 1,200
3 Malay sun bears	150	450
10 Exotic pheasants	100	1,000
10 Peacocks	10	100
5 Ringtail lemurs	200	1,000
5 Black buck	400	2,000
6 Iguanas	30	180
3 Orangutans	1,000	3,000
10 Gibbons	100	1,000
10 Lion monkeys	100	1,000
5 Proboscis monkeys	200	1,000
6 Langurs, capped	100	600
5 Cockatoos	100	500
6 Snub-nosed monkeys	100	600
87		$13,630

Australian Outback Exhibits

Select 300 to 900 acres of open pasture with a few dense hammocks or woods. South Texas, Arizona, New Mexico or Southern California would be ideal. But it could be done in Florida.

Australian Stock

Specimen	Cost per specimen	Total cost
10 Kangaroos	$500	$5,000
3 Cassowaries	400	1,200
5 Emus	200	1,000
10 Wallabies, roos	250	2,500
28		$9,700

The total stock if you wish to combine the various exhibits, having a separate section for each specific type of fauna, would be as follows:

Stock	Total cost
African	$ 98,880
South American	11,935
Indo-Pacific	13,930
Australian	9,700
	$134,445

The above would be for those who would start out on a relatively small scale and allow the stock to reproduce slowly as the project grows. A percentage of the profits could be used to purchase additional specimens each year as finances permit. The animal selection above would almost be lost on the several hundred acres of land (with the exception of the elephant, white rhino, and giraffe), but within ten years, if they reproduce well, small herds will fill the area. You must expect to lose some animals, so order more to anticipate replacements.

A much more ambitious project that would cost a great deal more is presented below. It would be spectacular.

African Jungle on Ten Thousand Acres or More

30 African elephants
30 Hippos
50 Zebras
30 Giraffes
10 Camels
30 Ostriches
20 Elands
50 Baboons, mixed species
20 Chimps
20 Mandrills
10 Drills
30 Impalas
30 Gazelles, Grant's, Thomson
30 Cape Buffalo
30 Hyenas, mixed species
30 Lions
100 Small mammals, jackals, fennec, etc.
10 Black leopards (these are sometimes available)
30 White rhino
20 Black rhino
10 Gerenuk
30 Wildebeests
30 Giant Tortoises, Aldabra
30 Kudus
30 Gemsboks
10 Gorillas
25 Cheetahs
10 Pigmy hippos
100 Antelope, assorted
500 Monkeys, mixed species
500 Birds, mixed species
500 Flamingos
500 Reptiles, crocodiles, lizards, etc.

Total cost approximately $950,000

The above specimens would not all be acquired at one time, since it sometimes takes months or years to obtain certain species. Therefore, they would not have to be paid for all at once. I did not itemize the individual costs of specimens, for with a large project, particularly one that would have a permanent big-game status, it would most likely be possible to get many of the animals at a very low price, since some of them are likely to be surplus in African game parks and will be shot if there is no live requirement for the animals. Rather than kill the animals, the parks may let you have the animals free with the understanding that in some future year, they may need some of the animals to replenish their own stock. You would only need to pay the freight on the animals, plus the cost of handling and feeding, to their destination. Depending upon inflation and rising costs, the entire complement of animals, birds, and reptiles for a fairly large project would run between $950,000 and $1,500,000, but it could be a lot less if you were to obtain at least some of the animals as surplus from game preserves. The cost could be shaved further by eliminating a good portion of the birds, including the flamingos, but *not* the ostriches. You will need these huge birds for show. In Southern Florida, you will already have thousands or millions of birds in a wildlife hammock area. Many of the birds, such as the herons, egrets, anhingas, and pelicans, are also found in Africa, so the bird population of your exhibit will not be lacking if you provide good water and surplus food. The areas that I visited as possible sites for a jungle project in Central Florida had enormous flocks of birds of all types, including huge vultures, or buzzards. There were also wild boar in the area and huge alligators, so there was already a pretty good start in wildlife before adding a thing.

Specimens for a South American Rain Forest on Ten Thousand Acres

1,000	Monkeys, all species including marmosets
1,000	Caymans, assorted sizes
100	Giant anacondas
30	Jaguars
30	Ocelots
50	Giant anteaters
30	Sloths, two species
10	Giant otters
50	Tapirs
100	Capybaras
30	Rheas
1,000	Birds, macaws, toucans, parrots, etc.
500	Iguanas
1,000	Small reptiles and animals, turtles, etc.
20	Giant armadillos

Total cost approximately $150,000

This could be a fabulous wildlife preserve. It would require choice, heavy hammock land that could be found in Central or South Florida, preferably the latter. The land would need to be enhanced with many more tropical trees and vines. Rich soil would need to be brought in, and a constant supply of fresh water would be an absolute necessity. South America has many unusual monkeys rarely if ever seen in zoos. To see them out in the wild in such a jungle area would be an exciting and interesting experience. Guests could visit the area in authentic dugout canoes, either poled along by an "Indian" guide or motor driven by silent electric outboard motor so that the sounds of the many tropical birds could be heard. A thousand or more caymans, which are quite inexpensive, or a good supply of native alligators would add interest to the cruise, for the banks would be lined with them. The water in the area would have to be warm or they would freeze in the winter months. Pool-type oil heaters could be set in place in certain areas to be fired up in cold weather. In addition, deep water wells could be turned on to warm at least a portion of the water. Surrounding trees could be sprayed with this warm water to give the area a greenhouse effect and help to keep it warm. It would be foolish to simply dump all of the specimens into the area as soon as the fences or surrounding walls were in place. As with all new projects, go very slowly for the first few years, introducing a small amount of fauna in the spring so that it will have summer and fall to become accustomed to our weather. Only dense, tropical hammock of hardwood trees should be considered. Pineland would be entirely unsuitable for such a project. The farther south, the better for such a project.

There are some beautiful junglelike forests in the Silver Springs area, and already both squirrel monkeys and the larger rhesus monkeys are living there out in the wild, having survived the cold winter in the area for many years. I first saw the rhesus monkeys there in 1943, and they are still living in the wild to this date, having reproduced somewhat through the years. I do not know how long the squirrel monkeys have been living there, but it has been at least five to ten years. This proves that the monkeys can adjust to our colder climate. They would certainly survive in Southern Florida as they do in the Monkey Jungle near Miami, where many are reproducing.

In correspondence with the famous scientist and naturalist Ivan T. Sanderson, it was pointed out to me that a number of years ago, a pharmaceutical company in the New Jersey area watered a large tract of hardwood forest with overhead sprinklers on a daily basis year round. The trees grew to unprecedented heights. If the trees would grow that well in the cold North, a similar project in Florida could produce a true South American rain forest. This could be a most exciting venture.

Summary of Basic Problems

The biggest problem in building either an African or South American jungle-type exhibit would be the acquisition of the land. Next would come fencing. This could cost as much or more than the land. Chain link would probably prove to be the most practical material, although for special purposes, it might prove worthwhile to investigate the cost of a cement wall versus the cost of chain link. A smooth cement wall, eight or ten feet high and sunk into the ground an additional four feet, would confine just about everything with the exception of elephants, leopards, and a few other species. It would also keep native animals out, which would help in the control of disease. It would also confine snakes and other reptiles that could simply crawl through a fence when they are young.

As to whether such a venture is worthwhile, as a tourist attraction alone, if the project is fairly large and if it is done in a natural way so that no fences are visible, the animals are healthy, and the area is of sufficient size so that the animals will not eat up all of the natural vegetation, the project could realize a profit the very first year, with proper management. Consider the project in the same light as a cattle ranch, only instead of raising cows, you will be raising rhino, giraffes, zebras, and exotic antelope. Should you decide to sell any of these animals, they will be worth two to five or ten thousand dollars each, whereas a cow raised in the same area would be worth just two or three hundred dollars—quite a difference, when you consider that all eat the same kind of food. This is why I stress a large piece of land for an African jungle project. Nearly all of the spectacular animals are grazers or browsers and feed mainly on grass and shrubs. You may need to do supplementary feeding to keep the stock in good shape, especially during the dry seasons. This is all that would be needed in the way of food so far as the hoofed stock is concerned, depending upon your area and the type of animals you keep. A herd of elephants or hippos will need a huge amount of food each day, and you may find it necessary to hire farmers to grow high grasses or sugarcane in rotating areas to sustain the animals so that they will not eat every tree and bush in the area.

As for meat eaters like lions and cheetahs, you will have to provide meat for them and this can be very costly. In the case of the cheetah it would be worth it because they are valuable animals, but lions are not worth very much and they may be difficult to sell. You cannot keep all that are born because they breed quite rapidly, but you will need some of them, for no African jungle could be complete without the king of beasts. Economics usually is a factor, especially in a stock-financed project, so if feasible it is wise to select animals that are likely to be of value and in demand by zoos and other attractions. Rhino are nearly always in short supply, as are giraffes, certain species of zebras, and nearly all of the African hoofed stock.

Many species of monkeys, both African and South American types, are quite valuable. Therefore if you investigate which species are in biggest demand by zoos and animal attractions, you might consider heavy stocking of these and a light stocking of the common varieties, if this is feasible. A good yardstick as to the value of an animal will be your total cost in acquiring it. Also, an investigation and evaluation of animal dealers' price lists, disregarding the fly-by-nights, will give you a good idea of the current price and availability of most animals and birds. You must keep in mind the rising costs in wages and freight, which will increase the cost of all animals that must be imported. The freight alone can amount to several thousand dollars on heavy animals like rhino and elephant, so raising them here could be well worthwhile.

Therefore, in addition to having an attraction, an outdoor zoo or jungle preserve can become a lucrative source of income just from the raising of the animals alone. In fact, without even letting one person through your gates, if your project is on a large enough piece of land for rotation of the grazing areas, you could probably make enough on the sale of surplus animals alone to more than pay for the investment. The admission of the public would be gravy. In working out plans for my own project, I contacted a number of zoo directors, asking their advice about the feasibility of such a venture. Most were highly enthusiastic, including the well-known Marlin Perkins of the St. Louis Zoo. The consensus of opinion was that most zoos would welcome a source of animals within the country. This alone would justify the building of large animal preserves, for as our country grows, more zoos will be constructed and they will need animals, not to mention our good neighbors in Central and South America who are also building zoos. With the new trend to the large, open-air zoos, animals are likely to be needed in dozen lots rather than in pairs, for a dozen giraffes in a veldtlike setting is certainly more spectacular than just two animals.

But the main purpose of an animal preserve or big-game jungle attraction should not be for the sale of animals. This would only be for surplus stock or exchange of excess animals for new species. The whole concept of the venture should be for the conservation and propagation of wildlife. It should be open to the public so that they can see the animals, photograph them, and enjoy them. It should not be a temporary project where the animals become acclimated to the land and then it is abandoned to the bulldozer and housing lots. This would be a terrible loss to the person or persons who originated the idea and put their heart and sweat into making it a reality. Nothing is sadder than to see an animal exhibit close and tear down its fences. The animals suffer the most, for they must be rounded up and disposed of quickly, usually at a great loss of life, when they are taken from their homes so quickly. If an animal preserve and jungle exhibit is to be established, it should be a permanent thing or it should not be started at all. It would make little sense to go through all the expense and forethought of bringing animals from Africa and other nations and acclimating them to an area and then, just as the animals become adjusted and secure, find it necessary to abandon the area to some land developer. This would be horrible. The land should be secured as a wildlife park, with a permanent status as such. Those who start it should not abandon their charges. For this reason, it is best to start small so that you can see how much the animals will eat and how much it is going to cost you to maintain them on a year round basis—that is, of course, unless you have a great deal of money. Then you could start off any way you like. The main thing is to acquire a large piece of land. Start out small by stocking just a portion of it at a time so that you will not get in over your head.

You will find many of the most desirable animals on the new endangered species list, which will mean that you may not be able to obtain these animals at all, except with a great deal of effort. As with all laws, there are exceptions and special permits attached, so should your project be large enough to provide a secure and permanent home for some of the animals, then you should be permitted to bring in a few of them for propagation purposes. This is actually insurance that the animals will survive, for many of the endangered animals are found in very restricted areas where an increase in human population, a small war, or a natural disaster could destroy the species forever. A typical example is the Celebes black ape. This is one of the most fascinating animals I have ever seen. It has a crest of hair on its head that it can erect at will and is a most interesting animal. But now it is on the endangered species list, which means that you would have difficulty importing it. However, I feel that those who enforce these regulations would consider special importation of these and other endangered animals if they are to be kept in an outdoor zoo or jungle in natural conditions.

Do not expect to buy or lease a huge tract of land and start stocking it with animals. It does not work out quite that way. You will find all kinds of obstacles, from politicians to zoning regulations, with which to contend. You will need a public relations department to convince local people that they will not be eaten when you turn your lions loose. Cattlemen will regard you like the plague. They will fear for the safety of their herds, and with good reason. But you can dispel their fears with your plans for strong, double fences or a wall. The double fences should be thirty feet apart from one another at the outside perimeter of your land. This is to guard against ticks crawling from your compound to any public land. You can mention that your animals will be government inspected for disease and that you will have a veterinarian in your employ. The local chamber of commerce will probably welcome your project, for it will bring fame and recognition to the area, not to mention the important tourists and new business for local motels.

You will need farmers, horticulturists, ranchers, veterinarians, and parasitologists, as well as experienced animal handlers. You may have to import help from the far reaches of our country or even from Africa, depending upon the size and scope of the project. We do not have all the answers to keeping animals outdoors. No one can positively say that a certain size fence will confine all animals. One zoo had a case where an elephant jumped over a high wall, something unheard of with elephants. There is still a great deal to learn about confining animals for the outdoor zoo or jungle. But we have learned a lot and by visiting the already established exhibits and by reading animal journals, we can share the knowledge of what has been learned and perhaps contribute something ourselves.

America could use many large outdoor zoos and African or Asian big-game preserves. How it will be done remains to be seen. Perhaps one way would be to form an amalgamated wild-animal propagation corporation that would be open to the general public, all zoos, animal importers, and all animal attractions. Membership fees and stock could be sold, and in this way, enough funds could be accumulated to build several large projects in various parts of the country. In this way, the people who are the most interested and who have something to gain would be supporting it. Rules and regulations would be enacted to perpetuate any projects started, and a voting system of one vote per stockholder could keep the more affluent from taking control. The stocks would be low enough so that millions of people could join, and members would be entitled to reduced rates for themselves and their family as well as to a sharing of the profits. It would appear that practically every animal attraction and zoo in the country would join, for it would insure them of a permanent source of good animals as well as a place to keep their surplus. Stock in the corporation could be offered for animals supplied by individuals or zoos and attractions, and the company could put out a small monthly magazine about happenings at the various big-game preserves. This is just one way to do it. In a nation that can send a man all the way to the moon, transporting the jungles from Africa or South America should be a cinch. A giant jungle in America could be a fantastic thing. The millions of Americans who would never have a chance to see Africa by going there could have Africa brought to them. It can be done. Let's do it!

13
Special Hints for Outdoor Enclosures

No zoo is one hundred percent escape proof so far as its charges are concerned. We can only do the best job possible based on modern technology and maintain constant vigilance. Some animals are, of course, easier to confine than others, and a thorough knowledge of the animals' habits is necessary to be reasonably certain that they will not escape. However, animals kept in large, outdoor enclosures are not as likely to get away as compared to those kept in steel cages in a zoo, because they will usually be contented with their home, and even if they do somehow get out, they will usually try to get back in—that is, of course, if they are provided with proper food and living conditions.

The easiest animals to confine are the hoofed stock, which comprise the bulk of the animal kingdom. This would include all of the species of animals that are grazers or browsers with the exception of the elephant. African antelope, deer, giraffes, wildebeests, zebras, camels, kudus, gemsboks, impalas, bison, and all the rest of the group, except for rhino and hippo, can be turned loose in a vast open area by simply encircling the area with a strong, chain-link fence ten feet high, securely fastened to the fence posts. The posts should be heavy and anchored in cement. It must be remembered that if a herd of animals becomes frightened and charges the fence, it can easily knock the wire and posts down unless they are well constructed.

Listed below are special hints in keeping animals outdoors. I do not wish to give the impression that I have all of the answers, because I do not. You can find many more from our wildlife directors, our zoo keepers, and our animal importers, trappers, and handlers.

1. Always round off corners in large enclosures so that animals will not come to the end of the fence. An oval or round enclosure is best rather than a rectangular one.

2. Select the proper gauge wire for chain link fences. All chain link is not the same. It may be obtained in varying strengths, some strong enough to hold a charging rhino. Also, it is available with colorful plastic coating, which may be more rustproof than regular wire. Do not expect chain link to hold elephants. They will walk right on through it.

3. It is best to bury the fence three feet into the ground to avoid animals digging out and, more important, in. Many valuable animals have been killed because dogs have dug a hole under a fence. In some areas the fence may need to be four feet in the ground if the soil is soft.

4. An eight- to ten-foot fence above ground will usually confine nearly all animals except certain cats.

5. A double electric wire at the top of the fence is recommended where animals may climb. This can be on a Y frame to keep animals from climbing in as well as out.

6. A water moat close to a fence will keep animals from charging the fence at top speed, particularly if they are frightened.

7. Full-grown elephants are extremely difficult to confine, but they can be confined by using standard railroad rails. These are cut into fourteen-foot lengths and driven six feet into the ground. They should be anchored in cement if the ground is soft. Holes are drilled through the upper portion and one-inch steel cable is threaded through the holes. At Addo Elephant Park in South Africa, this type of fence has held full-

grown, wild elephants, even some that have tried to crash the fence to get at oranges on the other side. Used rail is not expensive, considering its usefulness as fence posts. Once in place, it will last a lifetime.

8. Many animals can be confined with a simple water moat. These include chimps, gorillas, and gibbons, but not monkeys. These will require an electric fence or a smooth, six-foot-high concrete wall rising out of the water. This will confine them and is also effective against baboons as well as many cats.

9. Most outdoor zoos confine their lions in cages at night to keep control of them. They simply drive or herd the animals to their night quarters. The animals learn quickly and at the end of the day they expect it. A special handout of food is given to them after they are locked up.

10. Before planning any outdoor zoo, it is wise to visit as many of the established projects as possible and observe the latest in animal keeping. Usually if you write to them before you make your plans, they will be happy to show you their entire operation and may help you with your specific problems or questions. They may also have surplus stock that you may purchase and will of course expect you to keep them in mind when you have extra animals that they may need. Working together, a greater contribution to animal care is accomplished. Most outdoor zoos will not look on you as competition unless you are in the same county, since people today tend to visit the attraction closest to them.

11. Hoofed stock may be some of the most difficult animals to obtain due to strict rules of the government. In the case of African antelope, you may have to take only zoo-bred animals or animals that have been born in our country. Consequently, you may only get males, because zoo keepers are reluctant to sell the females.

12. White rhino are easier to obtain than black, even though they are much rarer. Compared to the black rhino, they are quite docile. They have reproduced so well in their native game preserves that they are in surplus at present. Many are being relocated to other game parks in Africa where they had been killed off in the past. Of course the situation can change and they could become very difficult to obtain, in which case black rhino would probably be more readily available.

13. The most difficult rhino to obtain are the Indian rhino. They have the most prehistoric appearance of all rhino. Only seven hundred remain in all the world. But, if an open-air zoo had a well-watered jungle retreat in Florida, Texas, or California and could assure positive protection of the rare animals, it is always possible that a few of them could be imported from India. Despite full government protection, the rare rhino are still being killed by poachers, and if a small colony were released in a protected area in America, it could well mean the survival of the species.

14. Do not neglect the hippo in your outdoor zoo. It is a fabulous animal and a great photographic subject. Furthermore, it reproduces well and is very hardy. A herd of twenty-five or fifty would make an exciting exhibit, particularly in a large lake where they could be viewed from a boat. Surplus hippos can nearly always be sold due to the fact that there are so many new zoos and animal farms starting each year. But should they ever increase to the extent that they are too plentiful for zoo sale, there is always the pot. Hippo hams are said to be far superior to hog, and there is far more usable meat per total weight on a hippo than there is on a steer. Furthermore, hippo tusks alone are worth a considerable sum as ivory.

15. Do not neglect grazing land when selecting a site for your zoo. Dense woods and hammocks are fine for shade and habitats for monkeys as well as for aesthetic value, but most animals prefer grass or brush. In Africa the biggest concentrations of all game are in veldt areas of grass and small trees.

16. Grazing areas should be fenced in such a way that the game can be driven into adjacent areas when and if needed. This way you can spray or replant the area in grass and shrubs. It also cuts down the disease factor.

17. The ideal small enclosure for reptiles, lizards, turtles, and small mammals is to enclose a small patch of trees and shrubs with a circular wall of smooth concrete, three- to six-feet high, depending what is to be confined there. The smooth walls will keep the creatures from crawling out, and the trees and brush will provide a natural setting as well as a home for them. Such an enclosure need have no roof, which will allow insects to fall into it, providing some natural food. In the case of old-world chameleons and other insect-eating lizards, a chunk of meat in the pit will draw a quantity of flies and other food for

them. This type of enclosure is far superior to any cage and requires little or no cleaning and maintenance.

18. An ideal birdcage may be made in a similar manner except that it must have a roof. For larger birds, select a small hammock and encircle it with surplus telephone poles. Then cover the area with wire, preferably small mesh, to keep out rats and snakes as well as to keep in smaller birds. A few poles in the center will keep the wire off of the treetops. Vines can be grown to cover most of the wire.

19. The real potential for open-air zoos, in addition to their conservation principles, is that many species will breed and reproduce much more readily in a natural habitat than they will in the more traditional zoos. If animals or birds become too numerous, excess may be trapped alive and either sold or exchanged for other specimens.

20. A parasitologist who is trained to work with wild animals would be invaluable for the open-air zoo, even more so than a veterinarian, although both would be best if finances could afford them. Depending upon the location of the project, neither would have to be on the payroll if they were both located in the area and could be called upon when needed. An experienced animal handler could take care of most of the problems except where broken limbs or disease are concerned.

21. You cannot learn too much about animals. Before you get too involved with an animal project, read a variety of books on animals, particularly books on keeping and handling them. (See "References and Suggested Reading" at the back of this book for a partial list of excellent reading on the subject.)

22. Northern climates should take advantage of the new styrofoam insulation. It would appear that very large, single-floor buildings with dirt floors and numerous skylights in the roof could be constructed with foot-thick styrofoam insulation and solar heating. These could be planted with tropical growth, freshwater streams, and rock outcrops, and stocked with exotic birds or small mammals. It would be, in effect, a greenhouse-type zoo and could prove very profitable and popular, especially in the frozen north.

23. Buy a jungle! Our wildlife is diminishing, not only from hunting, trapping, and pollution, but mostly from the destruction of the natural habitat, the forest and woodlands of our wilderness areas. Why not get together with a group of people who appreciate wildlife, such as a zoological society or any other group to which you belong, and buy a piece of the jungle? It can be done. You could buy a ten-thousand-acre or twenty-five-square-mile section of jungle in the name of your society and designate it as a permanent nature preserve with total protection of all wildlife within it. Then your society could visit the area on vacation, to inspect it and even patrol it if necessary against illegal hunting or development. You could perhaps buy it through an organization such as the World Wildlife Fund or through your own society. By yourself it would be a difficult thing to do, but with a society of several hundred members, in all likelihood it could be done. It would be a great contribution to wildlife conservation. You could buy it in the Amazon, Colombia, Ecuador, even Africa, or the United States.

24. Use extreme caution in releasing new animals, particularly antelope, into a new compound. Many valuable animals have been lost when they rushed from their crate and crashed headlong into the opposite end of the compound, breaking their neck. If the animals have been crated for a considerable time, they should first be released into a small enclosure with well-padded walls and as little excitement as possible, until they have settled down. Then they can be released into the larger area, which should be adjoining the smaller paddock. Also, a new animal added to a herd can cause friction and serious fighting. It is best to add the new animal when the other animals are being driven into a different section of the preserve. In the confusion it is less likely to be bothered.

25. Go on a wildlife safari to Africa and visit the huge wildlife game preserves. If you go with a group, it is safe and not too expensive. You will learn a lot about animals in the wild and also about exhibiting them in a natural area. In some places chain-link fence is simply placed around the area to keep the animals in, and that is basically all there is to an open-air zoo. The animals take care of themselves. You do not need any cage cleaners, zoo attendants, policemen, or signs. The only maintenance is for fence repair. The famous Kruger Park in South Africa is now largely fenced, but you cannot see the fences once you get inside the main gate. Nairobi National Park is just four miles from the center of a huge city, yet it has numerous lions and all the big game

except elephants. It has a fence at the city's edge to keep the animals off the streets. Yet no one is ever eaten or attacked by animals there unless he tries to approach the animals on foot, which is strictly forbidden.

26. If you feel that legislation regarding the keeping of exotic animals in your state is unfair, write to your governor and to Washington about it. Get your local zoological societies to do the same. Otherwise, we may soon see the day when only public zoos will be allowed to import or keep any wild animals. Individuals and the smaller private zoos will be shut out entirely, and this of course is unfair.

27. For really large animal parks, particularly public lands, it may be practical to dig a deep trench with a bulldozer around the entire piece of land, which would confine all of the animals, even the elephants. The inside edge of the moat could be scraped away so that animals could get back out of the ditch should they walk into it. The outer edge could be fortified with a good fence. This would contain both large and small mammals. It is used effectively in many of Africa's national parks.

28. Regarding the introduction of exotic animals to our land, it should be pointed out to opponents that man has introduced cows, horses, and sheep, none of which are native to this country. Have the wild burros or wild horses destroyed our ecology? It is a fact that they have not—but man has.

Appendix A

New Mexico's Experiment in Ecology
International Conference on Game Conservation and Wildlife Management
Palais des Congres
Monte Carlo, Monaco
December, 1968

BY

William S. Huey, Chief of Game Management
New Mexico Department of Game and Fish

In less than a month, North American hunters in New Mexico will begin a hunt for big game which one would expect to find only in a zoo, in a private park, or in its native habitat in North Africa. This hunt, along with others which have preceded it, is the result of an experiment in ecology which began in New Mexico almost 20 years ago. In 1950, 52 Barbary sheep or aoudad, *Ammotragus lervia,* were released into the Canadian River Canyon of east-central New Mexico. By 1955 their numbers had increased to the point that the first open season was held, and by 1962, over 1,000 hunters had been privileged to experience the unique thrill of hunting a North African big game species under wild conditions without the expense and trouble of a trip to North Africa. The results of this original experiment indicated that the scope of the project could be broadened with minimum possibility of adverse effects and the potential of significant benefit.

Research and observations conducted along with this first experiment have answered a number of questions and presented additional topics for consideration.

Much has been written and more said of the catastrophes which may follow the introduction of a nonindigenous form into a hospitable habitat. To illustrate the accuracy of these dire forecasts, we in America are reminded of the house sparrow, *Passer domesticus,* and the carp, *Cyprinus carpio,* among others. It is possible to rationalize that adverse effects of these species were the result of mechanical and economic factors that would have created problems from some of our native species had not the exotics been introduced. Accelerated agricultural practices are probably responsible for the tremendous increases in the numbers of blackbirds, *Agelaius sp.,* which now exist at the expense not only of the farmer but of some of the less dominant passerine forms. Unwise timbering practices along mountain streams have adversely affected water quality so that native trout, *Salmo sp.,* have been replaced to a large degree by native suckers, *Catostomus sp.*

As we consider the future of our experiment in ecology we realize that biological dangers do exist, but our experience so far with the aoudad tells us that with this species at least, and under the conditions which now exist, there is no adverse competition with either domestic livestock or native wildlife. We have demonstrated the possibility that should conditions warrant, severe control or annihilation could be accomplished.

Acceptance of the nonnative game animal by hunters has also been evaluated during the past Barbary sheep seasons, since the acceptability of a foreign animal by the average big game hunter was an unknown factor. From our experience we have found that the aoudad has been well accepted, and in spite of high fees and complicated application procedures, no difficulty has been experienced in fully subscribing the number of permits authorized.

In addition to consideration of the adversities, we also of course considered the many possible benefits. Not only does the introduction of additional species provide added recreational opportunity, it is also possible that some species doomed to extinction in its native habitat could be preserved and perpetuated here.

In July of 1961, the New Mexico State Game Commission formulated an official policy which expressed interest in the introduction of additional nonnative mammals into the more arid habitat of southern New Mexico.

The gemsbok, or oryx, with its magnificent horns, appears to be adapting well in New Mexico. There may come a time when the gemsbok will become extinct in its own land, and the transplanted stock in America will save the species. Surplus stock could eventually be shipped back to Africa. This has already been done with blackbuck antelope in Texas. It seems that the species thrived in Texas but was entirely killed off by hunters in its native land. Courtesy New Mexico Department of Fish and Game.

The oryx at Red Rock enclosure. The animals settled down well, and a fair number are now living wild in New Mexico, a milestone in broad-minded wildlife conservation. Courtesy New Mexico Department of Fish and Game.

Oryx at Red Rock Pasture. Some of these handsome animals have already been released upon public lands at White Sands. The author points out that cows, horses, goats, and sheep have all been released upon public lands without harming the ecological balance. Why not introduce African animals to select areas of our land? If they were to become too numerous, they could easily be shot by hunters. Courtesy New Mexico Department of Fish and Game.

Oryx enjoying their freedom at Red Rock enclosure. After a period of acclimation here, the young animals may be released on public lands. Courtesy New Mexico Department of Fish and Game.

proof enclosure was constructed at Red Rock, in southwestern New Mexico in 1964. This pasture which encloses approximately 320 acres has been used as a study area for a continuing research project conducted by New Mexico State University to determine the ecological effect of these imported animals on New Mexico's native wildlife and domestic livestock.

Our current inventory of Siberian ibex consists of 13 animals. During the five years since their arrival, our four surviving imported gemsbok females have produced 20 offspring, two of which have been lost through accidents. Fourteen are now being held at the Red Rock big game pasture, and four at the Albuquerque Zoo. Four of the zoo-born females now at Red Rock have produced six young.

Our current population of gemsbok include ten at the Albuquerque Zoo and 18 at Red Rock.

Of the 10 kudu imported, eight still remain at the Albuquerque Zoo, seven females and one male. Four young kudu were produced during late 1966 and early

This policy instructed the department to conduct surveys to determine what African and Asian species might be adaptable to our New Mexico situations.

By November of 1962, sufficient groundwork had been completed that the Commission authorized the purchase of Siberian ibex, *Capra ibex siberica,* gemsbok, *Oryx gazella gazella,* and kudu, *Tragelaphus* (*Trepsiceros*) *capensis.* Permission had been obtained from the U. S. Bureau of Land Management for the release of ibex into the Florida Mountains in south-central New Mexico, and the gemsbok and kudu were likely candidates for the mesquite grassland habitat type along our southern border. Along with the authorization for the purchase of these animals were provisions for the construction of pens at the Albuquerque Zoo in which the imported animals would be held. The U. S. Department of Agriculture requirements provide that imported animals may not be released into the wild, but must be held under approved conditions and only their offspring are eligible for release toward the establishment of wild herds.

The first of the imports, four Siberian ibex, arrived in Albuquerque in September of 1962. These were followed by seven gemsbok in August of 1963, and eight greater kudu in December of 1963. These imports were later supplemented by additional ibex and kudu. We had not anticipated that the problems of holding wild animals under zoo conditions would be as critical as they proved to be. These problems resulted in losses of imported animals as well as their offspring. Fortunately the reproductive rate of the zoo animals has been significant enough to offset losses, and the time when we will have sufficient numbers available to make initial releases into the wild is now within sight.

In order to provide space where animals produced at the zoo could be held and expanded for release, a game-

The Persian gazelle is another animal with which the New Mexico game department is working in its exotic animal program. The author feels that the placid white rhino, giraffe, zebra, and ostrich would also live out in the wild and add great interest to our southernmost wilderness areas, yet have little effect upon the ecology. Courtesy New Mexico Department of Fish and Game.

Releasing Iranian ibex in the Florida Mountains of New Mexico. The animals are reported to be doing well in their new home. Courtesy New Mexico Department of Fish and Game.

Siberian ibex and young at Red Rock Preserve. The animals are kept in this natural area to accustom them to the area and climate into which they will later be released. Courtesy New Mexico Department of Fish and Game.

A Siberian ibex at the Albuquerque Zoo. This one holds a world's record for long horns. Courtesy New Mexico Department of Fish and Game.

This beautiful natural area at Red Rock was fenced to provide an acclimation pasture for Iranian ibex and other animals. Courtesy New Mexico Department of Fish and Game.

1967. None of these have survived. One of these died shortly after birth, another was lost in the process of tranquilizing prior to shipment to Red Rock, the other two were moved to Red Rock, but did not survive extreme weather conditions during December of 1967. It remains to be seen whether subsequent generations of this species will become adapted to New Mexico's temperatures. There are now nine kudu calves at the Albuquerque Zoo. These animals will be transferred to Red Rock in the spring of 1969.

One male and two female Iranian ibex, *Capra aegagrus,* were imported during the summer of 1965 as gifts of the Governor of the Province of Kerminsha in southern Iran to Governor Jack Campbell of New Mexico. These were supplemented by five additional females imported as gifts to the Department of Game and Fish by the Shikar-Safari Club during the summers of 1967 and 1968. Fifteen young have been produced from these females. Of these, 13 survive, 11 at Red Rock and two at the Albuquerque Zoo.

At the time of the importation of the Iranian ibex, two Elburz red sheep, *Ovis ammon* subsp. were received. This young pair has since produced two young, one of which survived. Although they are extremely handsome animals, it is doubtful that they will be found suitable for release into the wild due to the possible conflicts which might result in adverse effects on our native bighorn sheep populations.

During the summer of 1967 the Shikar-Safari Club also financed the importation of five Persian gazelles, *Gazella subgutturosa*. This importation was supplemented this year by two additional females. Our current inventory includes two males and six females. We believe there is significant possibility that these animals might be able to maintain the mobility necessary to take advantage of local moisture conditions in our arid shortgrass prairie regions which are withheld from our native pronghorn by their reluctance to cross sheep-tight fencing. During the summer of 1968 the Southern California Safari Club imported a herd of Markhor, *Capra falconeri*, a male and three females, and there are now two female kids.

All of the animals with which we are currently working will be the subject of continuing research, prior to any releases into the wild. Foremost among the objectives is to investigate the possible ecological effect which they might have on native wildlife or domestic livestock. The existing big game pasture at Red Rock and the second of these pastures, financed by the Shikar-Safari Club, play an important part in this research and in allowing us to expand our available stock to plantable numbers.

A kudu at the Albuquerque Zoo. The animals are kept at the zoo, and only the offspring may be released on public lands. Just imagine this magnificent animal roaming free in the wilds. It may not be able to take the cold weather in New Mexico, according to report. Courtesy New Mexico Department of Fish and Game.

Our immediate plans for the release of additional species into the wild are definite with respect to an initial release of the Iranian ibex into the Florida Mountains in southern New Mexico within the next two or three years. Permission for this release has already been granted by the U. S. Bureau of Land Management.

Permission has been also granted by the U. S. Bureau of Land Management for the release of Siberian ibex in the Ladron Mountains in the central part of New Mexico. A request has also been submitted for permission to release the gemsbok into an area in south-central New Mexico and, although no positive indications have been received as to how this request will be treated, if a negative response is received, there is another suitable possibility for the gemsbok.

We realize that success with all of our candidate species is not likely. I am sure we will be satisfied and our sportsmen well served if only one or two of these species become adapted.

Report on Exotic Mammal Program to the Shaikar-Safari Club, January 8, 1971

Our current total inventory of animals is presented below in tabular form. This tabulation may be subject to deletions in the very near future. During the past week New Mexico has been subjected to some critical cold weather and during that period at least three kudu have died and two more are in bad condition. This contributes to our thinking that the kudu is probably not adaptable to New Mexico's weather conditions and will probably result in immediate plans for the deletion of greater kudu from our program. So much for the bad news.

On December 8 we released fifteen Iranian ibex into the Florida Mountains. This release was accomplished with our helicopter ferrying the crated ibex from our big game trailer to the top of the mountain. This release was completed with only minor problems which we have come to accept as routine.

Two of the adult female ibex released carried small radio transmitters on webbed collars. Monitoring of signals from these transmitters following the release has been accomplished by a student at New Mexico State University. This is a continuation of our contract research to determine the adaptability of these animals to New Mexico's habitat, and their expected ecological impact on native species. Reports from this student are that the ibex are still within the release area and appear to be doing very well.

With this release we still have twenty Iranian ibex at Red Rock and seven at the Albuquerque Zoo and plan to make follow-up releases into the Floridas next December.

In 1969 we made a release of gemsbok on to the White Sands Missile Range northeast of Las Cruces in southern New Mexico. This initial release has been supplemented.

The total number of animals released to date are six adult females, two yearling females, three yearling males and two adult males. Reports indicate that there were at least five calves born there this year and there is a report of one animal found dead by maintenance per-

Greater kudus at the Albuquerque Zoo. The animals were brought in for propagation purposes and eventual release upon public lands as a game animal in New Mexico. Courtesy New Mexico Department of Fish and Game.

sonnel on the area. Army personnel at White Sands have been very cooperative and their attitude toward this release is such that we will probably be successful in establishing the gemsbok into this choice area.

Early in 1970 the third experimental pasture at Red Rock was completed with the financial assistance of a local sportsmen's group. It appears that the idea of the Shaikar-Safari Club in the construction of our Iranian ibex pasture has caught on among the local groups and, as a consequence, their support of the program has been much more enthusiastic than in the past. This new pasture, although smaller than the Iranian ibex pasture, is ideally suited to the Siberian ibex which we had previously kept in the low pasture along with the gemsbok. The Siberian ibex are doing much better in this new pasture and we have hopes that we can now begin to develop a number of these animals suitable for a release in the not too distant future.

Appendix B

Florida's Exotic Game Program

STATE OF FLORIDA
Office of Governor Claude R. Kirk, Jr.

November 13, 1970

Mr. Robert P. L. Straughan
Salt Water Aquarium Magazine
and Coral Reef Exhibits
Post Office Box 1005
Miami, Florida 33156

Dear Mr. Straughan:

Thank you for your letter of September 22 reflecting your experience and thoughts about exotic game animals in Florida. I certainly agree that there are numerous animals which might be biologically suitable for our climate and habitat conditions.

Because of this, I am advised by our ecologists to be especially careful which species are released in our state. At the present time, the policy of the Game and Fresh Water Fish Commission calls for the careful screening and experimental introduction of a few animals to meet special sporting needs. Presently, they have under study the sambar deer of India (which happens to be available from stock on an island in Florida). The sitatunga and several of the waterbucks, both mentioned by you, are under consideration for some of our marshes which are not quite suited to our native deer. I am sure that the Commission would appreciate the advantage of your experience if you could find the time to send any information you might have.

Besides big game, the Game and Fresh Water Fish Commission is experimenting with the spotted tinamou of South America, the black francolin of India, and four species of ducks from South America. These species have been studied in their native habitats and considerable knowledge has been collected suggesting their adaptability to Florida conditions. Our game biologists believe that these species will survive in habitats not presently occupied by native animals with which they might compete and that they will be helpful in accommodating the ever increasing hunting demand.

As far as our biologists have been able to ascertain, the pronghorn antelope experiment was not successful. They believe that no antelopes involved in the experiment survive at this time. The thought has been expressed that the antelope should not have been selected in the first place.

For further information, feel free to contact our biologists who are in charge of this program at our Game and Fresh Water Fish Commission, Wildlife Research Projects Office, Gainesville, 32601.

Sincerely,
Claude Kirk
Governor

References and Suggested Reading

Books

Adamson, Joy. *Forever Free.* New York: Harcourt Brace Jovanovich, 1962.

Bartlett, Jen and Des. *Nature's Paradise.* Boston: Houghton Mifflin Co., 1967.

Brown, Leslie. *Africa: A Natural History.* New York: Random House, 1965.

Carrington, Richard. *Great National Parks of the World.* New York: Random House, 1967.

Crandall, Lee S. *Management of Wild Mammals in Captivity.* Chicago: University of Chicago Press, 1964.

Grzimek, Bernhard and Michael. *Serengeti Shall Not Die.* New York: Dutton, 1961.

Hallet, Jean-Pierre. *Animal Kitábu.* New York: Random House, 1967.

Leakey, Louis S. B. *Animals of East Africa.* Washington, D.C.: National Geographic, 1969.

Prince Philip, Duke of Edinburgh, and Fisher, James. *Wildlife Crisis.* New York: Cowles Book Co., 1970.

Sanderson, Ivan T. *Book of Great Jungles.* New York: Julian Messner, 1965.

———. *Living Mammals of the World.* Garden City, N.Y.: Doubleday & Co., 1961.

Schmidt, Karl P., and Inger, Robert F. *Living Reptiles of the World.* New York: Hanover House, 1957.

Spinage, C. A. *Animals of East Africa.* Boston: Houghton Mifflin Co., 1963.

Straughan, Robert P. L. *Jet Safari to Africa.* Cranbury, N.J.: A. S. Barnes and Co., 1972.

Sources of Literature Related to Conservation

Animals. 21–22 Great Castle Street, London W1, England.

Environmental Defense Fund. 162 Old Town Rd., Setauket, N.Y. 11733.

International Turtle and Tortoise Society, Box 45555, Los Angeles, Calif. 90045.

International Wildlife, National Wildlife Federation. 1412 16th St. N.W., Washington, D.C. 20036.

National Audubon Society. 1130 Fifth Ave., New York, N.Y. 10028 (*Audubon*).

Nature Conservancy. 1800 Kent St., Suite 800, Arlington, Va. 22209.

New York Zoological Society. Bronx, N.Y. 10460 (*Animal Kingdom*).

Salt Water Aquarium Magazine. P. O. Box 1000. Belleview, Fla. 32620.

Simian Society of America. 510 Lowell St., Wakefield, Mass. 01880 (*Simian*).

World Wildlife Fund. 910 17th St. N.W., Washington, D.C. 20006.

Zoological Society of San Diego. Box 551, San Diego, Calif. 92112 (*Zoonooz*).

Index